T0331769

Optimization Tools for Logistics

Series Editor
Jean-Paul Bourrières

Optimization Tools
for Logistics

Jean-Michel Réveillac

First published 2015 in Great Britain and the United States by ISTE Press Ltd and Elsevier Ltd

ISTE Press Ltd
27-37 St George's Road
London SW19 4EU
UK

www.iste.co.uk

Elsevier Ltd
The Boulevard, Langford Lane
Kidlington, Oxford, OX5 1GB
UK

www.elsevier.com

Notices

Knowledge and best practice in this field are constantly changing. As new research and experience broaden our understanding, changes in research methods, professional practices, or medical treatment may become necessary.

Practitioners and researchers must always rely on their own experience and knowledge in evaluating and using any information, methods, compounds, or experiments described herein. In using such information or methods they should be mindful of their own safety and the safety of others, including parties for whom they have a professional responsibility.

To the fullest extent of the law, neither the Publisher nor the authors, contributors, or editors, assume any liability for any injury and/or damage to persons or property as a matter of products liability, negligence or otherwise, or from any use or operation of any methods, products, instructions, or ideas contained in the material herein.

For information on all our publications visit our website at http://store.elsevier.com/

British Library Cataloguing-in-Publication Data
A CIP record for this book is available from the British Library
Library of Congress Cataloging in Publication Data
A catalog record for this book is available from the Library of Congress
ISBN 978-1-78548-049-2

Printed and bound in the UK and US

Contents

Foreword

In a society, where the fundamental role played by transport only occurs to ordinary people when it does not work or works poorly, logistics is not subject to prejudice. Citizens do not know the ins and outs of how it works and in most cases barely know anything about it, understanding it to be something vaguely related to the transportation and movement of goods. Constituting a social representation about logistics would require developing not only a view about the geographic division of work, but also creating an awareness about the infinite consequences of the mass consumer system, described by E. Durkheim[1] as a total social phenomenon.

The term "logistics" covers the three fundamental dimensions of the flow of goods encountered by all companies involved with goods manufacturing: supply, flows within the enterprise – known as production management – and distribution. Any movement – near or far – of people or goods requires the consumption of space, time and energy. Understood as a general discipline for the management of flows, logistics is strange in that its effectiveness is inversely proportional to its visibility. The better it performs, the less it is seen with both local and global economy players only becoming interested in it when, for various reasons, it is unable to respond to the complex problems it is asked to solve.

---

1 Emile Durkheim, 1858–1917, French professor of philosophy, considered to be the founding father of modern sociology.

"Thought stems from action before becoming action once again,"[2] stated H. Wallon. If there is any universe where this quote is true, this is certainly logistics. Logistics is to transportation and movement what thought is to action. Though, as the section on the history of logistics in this book demonstrates, militaries have for centuries thought about displacing their armed corps, it was not until the second half of the last century and the banalization – in Western societies – of movement, the advent of a global division of work and a globalized mass consumer society that the rationalized organization of flows was imposed as a discipline vital for the productivity of firms. Named in accounts as an expense, most enterprises involved in providing goods still equate it with loss. It is high time that it becomes an asset.

Logistics is constituted from empirical knowledge which is modeled to offer tools to rationalize movement. Each logistics concept or model is immediately faced with the reality of being effective: change the shape of the packaging of your product and we immediately change the characteristics of the transportation. Change suppliers and we change supply chain and storage techniques. Create a new product and we have to remodel our own production chain.

We may face the task of delivering four packages to two different points. Easy! But we may face the task of organizing the rounds of 140 drivers each of whom have in their vehicle (some of which belong to our enterprise, some of which are rented and some of which are subcontracted) over 10 tons of goods divided into over 250 packages that are to be delivered to 50 different points. Some of these delivery points may belong to the same client and be several tens or hundreds of kilometers apart. And so we find ourselves confronted with a puzzle with over a million options. Cartesians would take the straight line and lose time, space and money. Adopting the view of Pascal and modeling management on uncertainty would be better. However, developing a systematic and complex vision of the reality of flows and of their necessity to be really effective requires going beyond the triviality of common sense.

Until the oil crises in 1970s, the demand for transport was increasing daily. Consequently, there was little concern about vehicle fill rate, the pollution caused, the global cost of stocks as long as the "trip" was profitable.

---

2 Henri Wallon, 1879–1962, French philosopher, psychologist and politician, "De l'acte à la pensée", 1942.

At the beginning of the 21st Century conditions changed. In the international context of sustainable transport, for reasons of productivity, traceability, quality etc., it is no longer possible to ignore flow optimization, to be unaware of logistics. Knowledge accumulated by logisticians has to be shared, distributed, made common knowledge so that transport can be seen as a fundamental part of sustainable development and an inescapable dimension of the profitability of production. A product can only be called complete when it is available to who needs it: making this product available, logistics is an integral and inescapable part of the industry.

Each and every trade of goods requires information to be exchanged between the players concerned. Nowadays, industry players have to find the best match between geospace – territories and infrastructure – and cyberspace – the virtual word of data exchange.

Logistics is everywhere all the time. Certainly not everything is logistics but there is logistics in everything. However, knowledge about logistics is currently only shared between specialists. The aim of this book is to give meaning to, explain the modeling that is the result of the knowledge accumulated by logisticians and to find real applications for it. Containing countless graphs and images, this book is not merely a collection of decision support techniques: its aim is to bring the knowledge to a wider audience, an audience beyond the experts. My friend Jean Michel RÉVEILLAC has in this book purposefully reduced the number of long mathematical demonstrations to make flow management modeling understandable and favor operational research.

Logistics seeks the best match between all the coercive systems involved with movement so as to transform it into accessibility. And this must no longer be improvised because ultimately, in enterprises, it is humans who decide the actions of other humans.

Pascal MAUNY
Director of the IUT of Chalon sur Saône
University of Burgundy
August 2015

About This Book

Countless books have been written about logistics, operational research, decision support, graph theory, dynamic programming and so on, but very few have brought together all these domains to present a synthetic view which distances itself from aspects of pure mathematics, though without neglecting it completely, while offering lots of practical exercises.

The majority of the tasks presented in this book can be worked out with a simple calculator, a sheet of paper and a pen or with classic bureaucratic tools such as spreadsheets, or more specialist tools such as project managers and flow simulators.

The techniques presented and their domains of use are many and I am sure that a student, logistician, developer, technician, engineer, computer scientist, decision maker and you, dear reader, will find unexpected practical applications for them in your professional and even personal life.

Target audience

This book is aimed at anyone facing logistical problems related to flow management, decision-making, route or round optimization, meeting an objective when faced with multiple constraints, the creation of dashboards and relevant simulations, etc.

The tasks in this book require a minimum level of mathematical knowledge: a second year student reading science or economics should not encounter any major difficulties. I have tried to keep things simple and get straight to the point when it comes to theory and not concern myself with long, seemingly unnecessary demonstrations.

For the practical exercises, on a microcomputer, a good knowledge of the operating system (path, folders and directories, files, names, extensions, copy, displacement, etc.) is essential.

Lots of the tasks use a spreadsheet and so you will need to be able to use the basic features of this type of software competently. A basic knowledge of how to use pivot table type data handling tools is also beneficial.

If you are a follower of the visual basic application (VBA) language or one of its equivalents, you will be able to fully understand, improve, enrich and create new solutions to certain problems.

Finally, if you know the foundations of database management systems and relational algebra then you will be completely at ease in all the domains we will cover.

The organization and content of the book

This book is divided into two broad parts which have different aims.

The introduction and Chapters 1–8 discuss theory.

They present in turn:

– an approach to logistics;

– an overview of operational research;

– the foundations of graph theory;

– calculating optimal routes;

– dynamic programming;

– planning and scheduling with PERT and MPM;

– calculating flows in a network;

– spanning trees and rounds;

– linear programming.

This first part contains the fundamental concepts required to understand the remaining chapters in the book. Countless examples are presented alongside the theory and each chapter closes with a series of exercises and their solutions.

Chapters 9–13 focus on calculating different exercises using different software, respectively:

– a presentation of the different software for O.R. and logistics;

– operational research with a spreadsheet;

– dashboards with a spreadsheet and pivot tables;

– scheduling and planning with a project manager;

– simulating computer flows.

Each chapter centers around a number of examples built using classic (such as spreadsheets) or more specific software.

The conclusion, as its name suggests, attempts to assess the current state of logistics, its tools and future development.

Appendices 1 and 2 provide some additional information. They detail, in turn:

– installing a solver in Microsoft Excel;

– the table of the normal centered reduced law.

To follow up what was presented above, the bibliography contains a list of internet sites.

There is also a glossary to explain certain acronyms and terminology specific to logistics and operational research.

Conventions

This book uses the following typographic conventions:

– *italics* is used for quotations, mathematical terms, comments, equations, expressions or variables or theorems presented in the theoretical and practical chapters in the examples and exercises;

– *(italics)* are terms in a foreign language;

– CAPITALS are used for names of windows, icons, buttons, folders or directories, menus or sub-menus. They are also used for the elements, options or commands in the window of a program.

– **bold** is used for important terms the first time they are used in the text. These terms are generally found in the glossary at the end of the book.

– The notes are indicated by the keyword: **NOTE**. They are used for important comments to add to the explanations already provided.

– `courier` font is used for lines of VBA code. These lines may end in the symbol ⏎, which indicates a soft return.

All the figures and tables have a key to help readers interpret them.

Vocabulary and definitions

As is the case for all techniques, optimization tools for logistics have their own vocabulary. These words, acronyms, abbreviations and proper nouns may not always be familiar to readers. Please refer to the aforementioned glossary.

Thanks

I would like to particularly thank the team at ISTE Press and my editor Chantal Ménascé who trusted me. I extend my thanks to my very dear friend, Pascal Mauny, Director of the IUT of Chalon en Saône and Lecturer at the University of Bourgogne, for his time, attention and listening skills and for writing the Foreword to the book.

I would finally like to thank my wife, Vanna, who has supported me throughout the time I was writing this book.

Introduction

I.1. What is logistics?

Answering this question is a difficult task. It is easier to describe its utility, or rather its utilities, and even its uses or roles, of which there are many, than to formulate a definition. Nevertheless, I will try, by referring to various sources. Let us start with a good reference tool, the dictionary.

French dictionary Larousse 2015: *"Set of methods and means relating to the organization of a service, an enterprise, etc. including storage, transport, packaging and sometimes supply."*

The definition is clear, but vague, so let us look elsewhere.

Encyclopaedia Universalis 2015: *"In current usage, the term logistics refers to activities involving the transportation of materials or goods whose domains of applications are primarily military or relating to the military, for example, humanitarian aid during conflicts or famine. However, though logistics remains one of the major components of the "art of war", it has since the late 1960s been increasingly practiced in the context of enterprise."*

I will not present all the definitions I found in different dictionaries and encyclopedias, but there is one point that must be made: whenever the term "logistics" is mentioned, another term immediately appears at its side, that is, "military".

Let us finish with two other definitions. First, the official definition provided by AFNOR[1] (standard X50-600): logistics is a function *"whose aim is to meet the expressed or latent needs in the best economic conditions for the enterprise and for a determined level of service. Needs are of an internal nature (supply of goods and services to ensure the operation of the enterprise) or an external nature (satisfying clients). Logistics calls upon a number of professions and know-how that contribute towards the management and control of physical flows and flows of information as well as methods"*. Second, the definition found in countless (particularly French) articles, which for my taste remains a bit simplistic as it focuses too much on distribution: *"Logistics is guaranteeing the availability of a product or service in good conditions, at the right time, to the right place, to the right client, at the right price."*

I.2. A history

The term "logistics" has military origins. It is thought that Baron Antoine Henri de Jomini[2], a great strategist who published a number of books in the early 19th Century, is its founding father. At that time, logistics consisted of moving armies over a terrain while ensuring they received provisions (food for soldiers and horses, supply of arms and ammunition and simultaneously the implementation of means of transport).

Figure I.1. *Antoine de Jomini (Source: Wikipédia)*

1 Association Française de Normalisation: official standardization organization representing France at the International Organization for Standardization (ISO) and the European Committee for Standardization (CEN).
2 A self-taught military man, Antoine de Jomini published a number of books including, among others, "Treatise on Grand Tactics", "Treatise on Grand Military Operations" and "The Art of War". He was considered the leading specialist in military strategy of his time. He worked with Marshal Ney, Napoleon, Alexandre I, Nicolas I and Napoleon III.

Over time logistics has developed and though the term appeared just three centuries ago, some of its techniques seem to have been around since antiquity (Julius Caesar mentioned in his writing the problems he faced getting provisions to his legions).

It was not actually until the 17th Century that the real foundations of modern military logistics emerged (the creation of supply stores, management of horse and vehicle parks, appearance of the service corps).

In the wars of 1870 and 1914, the railway made its entrance and fully participated in the military supply chain, particularly with regards to ammunition.

The war of 1939–1945 saw the motorization of armed forces that were able to travel quickly and across considerable distances, crossing the borders of different countries. The implementation of proper supply chains became an absolute necessity using all means of transport (road, rail, water and air). The multiplicity of transportation processes made a war expanding over an immense geographic area increasingly complex. What is more, a motorized army requires large volumes of fuel. Operation "Neptune", for which preparations were made for over two years, turned out to be a real logistical challenge for the military that became critically important. At this time, countless innovations emerged, including, among others, the creation or the arrangement of special means of transport, the design and use of packaging units (containers, palettes, wrapping, etc.) and the design of storage and distribution infrastructures.

During this progress, logistics would join up with mathematics via **operational research (OR)**. Scientists would enter the domain and determine by calculations, while taking into account certain constraints, the key values that would characterize the itineraries, the sizes of convoys, supply rounds, the size of warehouses, etc. Modern logistics was thus born.

As the decades passed, computer science entered the industrial world causing a real revolution. The first planning and scheduling techniques emerged (**PERT, MPM, CPM**[3], etc) and became digitized. The materials resources planning or manufacturing resources planning (**MRP method**),

---

3 "Program Evaluation and Review Technique" , "Metra Potential Method", "Critical Path Method".

specializing in combinatorial problems linked to supply management, manufacturing ranges and assembly ranges, combining compounds and components, were used increasingly widely. Its second version, **MRP II**, became the flagship technique of production management. Stock management and monitoring became increasingly accurate and it was even possible to predict future supplies by statistically organizing purchases and future client demands. Computers made everything possible, with calculations that could not hither to be worked out in a reasonable amount of time being resolved in several seconds.

A notion (that was nevertheless not new) assumed great importance: the **flow**.

The first oil crisis of 1973 and then the second of 1979 saw the emergence of new methods. Just in Time (**JIT**), pull or push flows arrived, copying the management systems used in Japan, particularly by Toyota. The main objective is cost reduction; staff in the production chain become players and they are surrounded by machines, automated systems and robots. Less and less work is done manually, machines build other machines as humans monitor, while the quality constraints governed by standards, such as the famous **ISO 9000**[4], are respected.

Quality management would lead enterprises to divide up the tasks they perform by following a set of completely standardized processes. Workshops would be reorganized and production would be incessantly improved and increasingly rationalized.

Everything becomes logistics or rather logistics is everywhere. There is no such thing as one logistic, but rather there are logistics: logistics of transport, production, distribution, supply, and so on.

Since 2000, the integration of flow management has enabled digitized simulations of all types of processes to be performed. The internet and local networks favored exchanges and got rid of the notions of distances and borders along the way. Machines, humans and systems communicate, exchange, dialogue, transfer data at exponential speeds and in exponential volumes. This is the era of globalization. Information has become the

---

4 Family of standards devoted to the management of quality and its correct application via the implementation of certification principles.

watchword and it is logistics that supports, creates, transforms and distributes it.

Nowadays, at least two logistics groups exist within companies: **logistics of services** and logistics of material goods. Both groups are processes that improve distribution, service and information networks to respond as fast as possible and at an optimal cost to an order placed by a buyer. Logistics groups can even go beyond this task by forecasting sales or purchase orders.

All these processes must integrate perfectly with the environment, whether it is human, technological or ecological, while generating a flow of information that will enrich the record of past or future transactions without ever neglecting financial constraints.

Over time logistics has become one of a number of keys to general strategy in the world of business. It became so present that the term **logistics strategy** was born, using a set of decision-making and tactical principles to make clients extremely satisfied.

I.3. New tools and new technologies

Since the advent of enterprise resource planning (**ERP**[5]), which really took off in the early 1990s, data within the enterprise's information system has been centralized to avoid redundancy, multiple input, the risk of error and simplify records monitoring.

Production knows its net requirements in components and materials, it can start and plan manufacturing orders by considering all resources, such as the workforce and machines; the scheduling of tasks is monitored in real time.

Buyers, accountants and financiers find something for them in the system; there are management modules capable of refining the accounting figures. The same can be said for the management of human resources.

Over the years, ERPs have integrated other blocks of functions, namely distribution, e-commerce and supply chain management (**SCM**), the key element of goods and services logistics.

---

5 Some of the most commonly known systems are SAP, MFG/Pro, Baan, SSA (Ex Baan), etc.

Nevertheless, ERP has reached its limits in recent years as its weight does not favor enterprises such as start-ups whose reactivity and changes to internal structures may be very fast. Moreover, their high costs and long installation and integration times do not work to their advantage.

Other tools have emerged, such as customer relation management (**CRM**), warehouse management system (**WMS**), supplier relationship management (**SRM**), knowledge management (**KM**), product lifecycle management (**PLM**), **project managers, digitized flow simulations**, etc.

I will stop here because the list goes on with new software entering the market on a daily basis. 2010 saw the explosion of a myriad of tools adapted and adaptable for all types of commercial businesses beyond the giants of ERP.

Nowadays logistics is a necessity that calls upon core sciences such as logic and mathematics via some of their branches such as the theory of postulates, Boole algebra, operational research, graph theory, etc.

1

Operational Research

1.1. A history

Like logistics, operational research (OR) emerged from a military context. The term is attributed to Watson-Watt[1] circa 1940 and became more widely known due to Patrick Blackett[2] who, during World War II, put together the first OR team to resolve air defense problems and later problems of getting provisions to troops.

Although not yet bearing the name, OR developed out of mathematics and problems relating to mathematical expectation or combinatorial analysis between the 17th and 19th Centuries. Blaise Pascal[3] worked on the problems of decision-making in uncertainty and Gaspard Monge[4], considered to be the father of optimization, studied the problems of cuts, fills and defilades. In the early 20th Century, stock management, with the famous Wilson formula[5], brought OR into the modern world.

---

1 Sir Robert Alexander Watson-Watt, (1892–1973), Scottish engineer specializing in radar.
2 Patrick Maynard Stuart Blackett, (1897–1974), physicist specializing in, among other subjects, nuclear physics. Inventor of the *"Wilson cloud chamber"*.
3 Blaise Pascal, (1623–1662), French mathematician, physicist, philosopher and theologian. In 1641, he invented the first calculator (the "Pascaline").
4 Gaspard Monge, (1746–1818), count of Péluse, French mathematician specializing in descriptive geometry, analytical geometry and infinitesimal analysis.
5 Also known as Economic Order Quantity, it calculates the optimal supply period of a production system. It was formulated by Harris in 1913 and realized by Wilson in 1934.

After World War II, large numbers of big enterprises started using OR, which was making sweeping progress. It would be taught at Massachusetts Institute of Technology (MIT) from 1948 before it spread to countless other universities and higher education institutions throughout the world.

Nowadays OR is known as a **decision support tool**.

1.2. Fields of application, principles and concepts

OR can be found within countless services in the enterprise. It is often invisible, like logistics, with which it works in close collaboration.

Its presence is, however, more particularly found in vertical applications such as scheduling and planning, production management, quality, purchase and supply, stock and storing management, conveyance, expedition and transport, commercial action, management control, human resource management, etc.

Methodologically, the use of OR is supported by well-defined principles resulting in the conceptualization of an often transversal approach through the enterprise's chain system.

The approach can be divided into a number of phases:

– identifying the problem;

– modeling the problem;

– solving the problem;

– validating the solution;

– implementing the solution;

– improving the solution.

COMMENT.– The next sections aim to clarify the general methodology to be implemented such as to formalize an OR problem as well as to define the specific vocabulary used by logisticians.

1.2.1. *Identification*

Identification is by far one of the most difficult phases. At this stage, one or more **objectives** and a set of **constraints** must be defined.

While at first glance this may seem simple, with a closer look it quickly takes on a complexity that often lies beneath the surface. For instance, in the case of the transportation of a load from one point to another, it could be said that the aim is to reach the town of arrival by spending the least amount possible on the delivery of merchandise. The constraints are the potential routes and their respective mileage, the choice of one or more suitable vehicles depending on the mass being transported, the consumption of these vehicles, the cost amortization per kilometer, etc. The delivery point in the town of arrival, however, may have specific opening hours, its storage capacity may be limited, etc.

It is clear to see how the objective itself becomes a source of constraints. It may perhaps need to be redefined with some of the previous constraints becoming in reality a new objective, such as, for instance, optimal transportation and delivery time.

Constraints are generally equalities or inequalities which constitute equation systems that are difficult to handle and solve. The objective is, in many cases, attached to a function for which a **maximization** or **minimization** is sought. This may also be the establishment and the verification of a relation.

There are often numerous and poorly defined criteria. The success of this initial phase – the identification of the problem – is crucial as it is where the future problem of the OR is formulated.

Since enterprises have been implementing quality management by modeling each of their key actions by one or a number of processes, logisticians have been able to draw on these resources. Nevertheless, it is rare for everything to be formalized enough such that objectives and constraints can be correctly defined. Only an accurate and detailed analysis, as well as collaboration with players on the ground, will provide the elements required for the problem in question to be correctly modeled.

1.2.2. *Modeling*

At this stage, the logistician will formulate an elementary description of the project by defining a set of variable integrated into equations, inequations, systems, functions or relations.

The nature of these variables can be either quantitative or qualitative.

All units (kg, l, g, m/s, h, m^2, km^2, m^3, °C, W, Kwh, etc.) are possible and in a diverse range of formats: integer, decimal, real, rational, monetary, hourly, logical, binary, personalized, etc.

They may be input, output or control variables.

Constant and random values may intervene.

A model is **deterministic** when the set of its parameters is known with certainty or **stochastic** when its parameters are uncertain.

Variables can represent the known or unknown values of a problem: they are often called **alternative**. They participate in **restrictions** that are developed and contained within constraints with a view to obtain a definite result or solution in the form of an **objective function** to be optimized.

A model is built based on a set of properties:

– **real properties** which belong to reality;

– **formal properties** which only belong to the model;

– **compatible properties** that adapt the model so that it fits with reality.

Depending on the chosen or imposed properties, logisticians are able to generate a **perfect model** when their model only contains the properties existing in the limited perimeter of the problem or a **complete model** when all the existing properties are taken into account.

The model can contain functions meeting a number or prerogatives:

– explaining the situation, present the causes and effects inherent to the problem;

– providing a solution enabling the variables and constraints of the problem to be acted upon such as to obtain a solution that gets closer to the optimum;

– integrating projected data to improve the solution.

To conclude this section about modeling, it can be said that building a model is based on hypotheses and a choice of mathematical tools.

Hypotheses are built by combining different principles:

– **Linearity** which enables doing sums with constants, variables and constants multiplied by variables or even multiplying or dividing constants between themselves. However, it prohibits the product or the quotient of two variables. Mathematically, linearity determines whether a system has a response close to a straight line.

– **Divisibility** used to find quotients between variables or variables and constants.

– **Convexity** implies that constraints are linear expressions.

– **Statistical independence** which gives perfect autonomy to a sequence of events (this is rarely the case in reality as causalities are often found between a number of events).

– **Stationarity** which presupposes no change during a defined time period.

– **Absence of memory** which consists of saying that a state obtained at the present moment is sufficient for determining a forecast independent of the past or future.

The mathematical tools that can be used to solve the problem are many and are borrowed from algebra, geometry and statistics. To name but a few:

– functions;

– integral calculation;

– linear equation systems;

– series;

– numeric sequence;

– matrices;

– graphs;

– Markov chains;

– and so on.

1.2.3. *Solution*

Solution methods are generally iterative, linking solution to solution until the optimum is reached. However, the latter is not always reached and in this event a satisfactory solution is deemed sufficient. In the first case, we talk of **exact** solution methods and in the second, we talk of **heuristic** solution methods.

Widely used methods employ the processes of **analytical solution, deterministic algorithms or simulation procedures**.

Graphs or algebraic methods may also be used depending on the number of variables and constraints that need to be processed. Graphs are often restrictive as they can only work within the plane or space and thereby limit the number of variables that can be processed.

Many algebraic methods arrive at the same solution.

Computer tools available today can process complex calculations in a very short space of time and approaches using successive approximations are possible. Computer tools are also the authority when a simulation is required.

1.2.4. *Validation*

In this phase, the obtained solution will be studied with regards to the constraints and objective to be reached. Two cases that may be validated can be defined as such:

– an **optimal solution**, which optimizes the objective function;

– an **acceptable solution** where a set of values found will meet all the constraints.

A logistician's job is not done merely when a solution has been found; the solution must be checked to ensure it corresponds to a reality true to the situation for which the model was defined.

If the result is not appropriate, a sampling error perhaps occurred in the variables used. The chosen relations, equations or statistical laws may not be adequate and so a detailed check takes place and the model must be looked at again for corrections.

1.2.5. *Implementation*

Once the solution has been validated, the method developed can be formalized and applied to other scenarios using similar constraints and the same objective.

The use of computer tools offers greater flexibility and quasi-limitless calculation power in terms of the implementation of different models.

To deal with these, logisticians can turn to spreadsheets (Microsoft Excel or others) or more specific software such as project managers (Microsoft Project, Sciforma, Microplanner, etc.) and flow simulators (ExtendSim, Flexsim, Arena, Witness, etc.).

These tools have several undeniable features:

– the option to work collaboratively;

– the option to work in a network;

– the option to exchange and publish online.

1.2.6. *Improvement*

When a method comes up with a solution to a given problem, it is not clear whether the solution could be improved on the technical level (or rather the mathematical level in our case) or the detail obtained in the results.

It can be interesting to analyze the decisions made by applying the results on the ground. What is their impact in conceptual or organizational terms? Could they be improved? Should other factors be taken into consideration? Is the optimum obtained unalterable? Have all contextual elements been included? Is it possible to rationalize the number of processes for which the operational model has been designed?

To conclude, like a doctor sitting in front of a patient, a diagnosis must be made so as to predict a justified and justifiable improvement of the model.

1.3. Basic models

Over the years, research in OR has created countless algorithms and diverse and varied methods adapted to precise fields of application encountered within all logistics systems.

In this book, the author has tried to bring together the main categories of problems encountered by adapted OR tools and their often very specific models:

- **linear programming**;
- **dynamic programming**;
- **optimal paths**;
- **scheduling**;
- **trees, rounds and transport**;
- **maximal flows and networks**.

The author has certainly not taken stock of everything: there are certainly other subjects which he has not discussed. However, in his opinion, this book deals with the most classic and current topics.

1.4. The future of OR

In increasingly complex environments, operational research and decision support (**ORDS**) is making itself more and more at home. It supports managers in making strategic choices which boost enterprises' competitiveness.

Though it remains the prerogative of specialists such as logisticians, some engineers and financiers, gradually, since the 1990s, OR has been entering commercial companies and industrial manufacturing after a long period spent in the academic sphere.

Countless tools and applications have emerged bringing OR to the reach of non-specialists.

In a context where globalization has become a belief, where competition is fierce, where stocks are expensive, where management decisions have become complex and where Just-In-Time (JIT) manufacturing tends to take over, OR is often essential and seems to have a bright future ahead.

Though in France, we are still behind our European neighbors and the United States, there has been a real breakthrough of ORDS in this second decade of the millennium.

2

Elements of Graph Theory

2.1. Graphs and representations

It is difficult to give an accurate date as to when graph theory was developed; however, work by Leonhard Euler[1] in the 18th Century is without doubt at the origin of research into numerous problems. Initially anchored in mathematics, it then developed within other disciplines such as biology, chemistry and then the social sciences.

In the early 20th Century, research by Menger[2], König[3], Berge[4] and others created an important domain in mathematics.

Graphs are everywhere in our everyday lives: in the road network, maps, computer networks, programming, electric circuits, medical diagnosis, animal biology, decision support, logistics, etc.

---

1 Leonhard Euler, 1707–1783, Swiss mathematician and physicist to whom we owe countless discoveries in the domains of infinitesimal calculus and graph theory. He also worked on the dynamics of fluids, optics, mechanics and astronomy.
2 Karl Menger, 1902–1985, Austrian mathematician specializing in geometry, game theory and graph theory.
3 Dénes König, 1884–1944, Hungarian mathematician, the author of the first manual on graph theory.
4 Claude Berge, 1926–2002, French mathematician and artist who contributed toward the development of graph theory. He specialized in combinatorial analysis. He was awarded the Euler Book prize in 1933. He also wrote literary books, particularly detective stories and short stories, as well as being a sculpture and an inventor of board games.

Countless problems can be modeled with a graph, a simple diagram containing vertices and edges.

2.2. Undirected graph

A graph G is formed of two sets $S=\{s_1, s_2, ..., s_n\}$ whose elements are called vertices and $A=\{a_1, a_2, ..., a_n\}$ whose elements are called edges (or lines), $G=(S, A)$.

An edge a of the set A is defined by a couple or an unordered pair of vertices which represent its endpoints. If the edge a connects the vertices x and y, these vertices are said to be **adjacent or incident to a.**

What we called the **order of the graph** n is the number of vertices it contains.

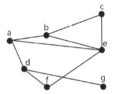

Figure 2.1. *A graph*

In Figure 2.1, S={a, b, c, d, e, f} and A={(a, b), (a, d), (a, e), (b, c), (b, e), (c, e), (d, f), (d, g), (e, f)}

The number of potential graphs is infinite and they can be divided into a number of categories.

2.2.1. *Multigraph*

A **multigraph** is a graph where one or more vertices can form a loop and/or multiples edges between two or more vertices.

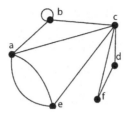

Figure 2.2. *A multigraph, the vertex b has a loop and a and e are connected by two edges*

2.2.2. *Planar and non-planar graph*

If a graph can be drawn without any edge being cut, it is said to be a **planar** graph.

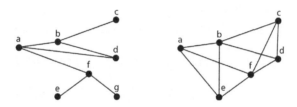

Figure 2.3. *A planar graph (left) and a non-planar graph (right)*

2.2.3. *Connected and unconnected graph*

A graph is **connected** if a vertex can be joined from any other vertex by following one or more edges. An *unconnected* graph can be broken down into a number of **connected components**.

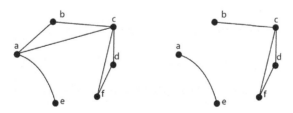

Figure 2.4. *A connected graph (left) and an unconnected graph with two connected components {a, e} and {b, c, d, f} (right)*

2.2.4. *Complete graph*

A graph is **complete** if each of its vertices is connected to all the others.

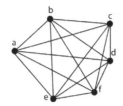

Figure 2.5. *A complete graph*

2.2.5. *Bipartite graph*

A graph is **bipartite** if its vertices can be divided into two subsets and each vertex of one subset is joined to each of the vertices of the other.

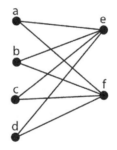

Figure 2.6. *A bipartite graph*

In Figure 2.6, the graph can be divided into two subsets of vertices $S_1=\{a, b, c, d\}$ and $S_2=\{e, f\}$. Each vertex in S_1 has edges going toward each of the vertices in S_2 and *vice versa*: the graph is bipartite.

2.2.6. *Partial graph, subgraph, clique and stable*

When one or more edges are removed from a graph G, we get graph G', which is known as the **partial graph** of G.

When one or more vertices are removed as well as their adjacent edges from a graph *G*, we get a *subgraph G'*.

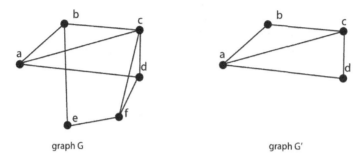

graph G   graph G'

Figure 2.7. *Graph G and its subgraph G'*

A partial graph of a subgraph is a **partial subgraph**.

A **clique** is a complete subgraph of G.

A **stable** is a subgraph of G with no edges, i.e. a set of vertices.

2.2.7. Degree of a vertex and a graph

The number of incident edges to a vertex s is called the **degree of the vertex** and denoted as d(s).

The degree of a graph is the maximal degree of the set of its vertices. When the graph has vertices which all have the same degree, it is said to be **regular**.

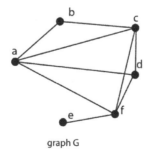

graph G

Figure 2.8. *A four degree graph (on a, c and f)*

In Figure 2.8, the degrees of the vertices are: $d(s_a)=4$, $d(s_b)=2$, $d(s_c)=4$, $d(s_d)=3$, $d(s_e)=1$, $d(s_f)=4$.

The total degree of the vertices of a graph is equal to two times the number of its edges. This is called the **handshaking lemma**.

This lemma is also valid in multigraphs with loops if one loop equals two loops when calculating the degree of a vertex.

2.2.8. *Chain and cycle in a graph*

A **chain** is a sequence of vertices and edges starting and finishing with a vertex. It has a **length** that corresponds to its number of edges. It is said to be *elementary* if each vertex is present just once and **simple** if each edge appears once at most.

A **closed chain** starts and finishes with the same vertex. It is only a *cycle* if the start vertex appears just twice.

The **distance** between two vertices is equal to the length of the smallest chain linking them.

The **diameter** of a graph is the longest distance between two vertices.

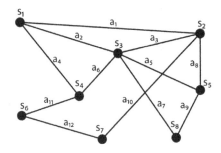

Figure 2.9. *A graph with 8 vertices and 12 edges*

Some examples from Figure 2.9:

– a chain: (s_1, a_2, s_3, a_5, s_5, a_5, s_3);

– an elementary chain: (s_6, a_{11}, s_4, a_6, s_3, a_3, s_2) with a length of 3;

– a closed chain: $(s_1, a_1, s_2, a_8, s_5, a_8, s_2, a_3, s_3, a_2, s_1)$;

– a cycle: $(s_1, a_2, s_3, a_3, s_2, a_1, s_1)$;

– the distance between s_1 and s_8 is also *2*;

– the diameter of the graph is *7*, let the chain be $(s_1, a_1, s_2, a_8, s_5, a_9, s_8, a_7, s_3, a_6, s_4, a_{11}, s_6, a_{12}, s_7)$.

2.2.9. *Level of connectivity (or beta index)*

The level of connectivity (or beta index) measures the density and the variety of the possible relations between the vertices of a graph, whether they are direct or indirect.

The *level of connectivity (or beta index)* b is the relation between the number of edges and the number of vertices.

$$b = \frac{m}{n}$$

where

m: number of edges

n: number of vertices

The **level of connectivity (or gamma index)** g is another version of the previous index. It is between 0 and 1.

$$g = \frac{m}{m_{max}}$$

where

m_{max}: maximum number of edges possible. If the graph is planar, this number is *3(n-2)*.

The **circuit rank (or cyclomatic number)** measures the maximum number of independent paths that can be simultaneously built within a graph.

$$v(G) = m - n + p$$

where

p: number of connected components (unconnected portions of the graph)

In Figure 2.9, $\nu(G) = 12 - 8 + 1 = 5$

2.2.10. *Eulerian graph*[5]

A *Eulerian* graph is a graph in which it is possible to find a cycle that passes once and only once through all the edges.

It can also be formulated in the following way. If $G=(S, A)$, it is Eulerian when for each vertex s of S, $d(S)$ is even. Indeed, if we consider G to be Eulerian, c is a Eulerian cycle and s is a vertex of G. The cycle c contains all the edges of G and as a consequence also all the $d(s)$ edges with s as an endpoint. When we follow c, we arrive at s **as** many times as we leave it, each edge of G being present just once in c, implying that $d(s)$ is equal.

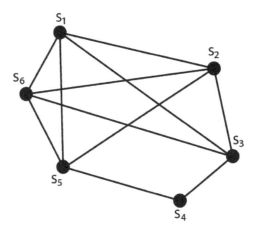

Figure 2.10. *A Eulerian graph*

The graph in Figure 2.10 is Eulerian as each of its vertices is even (s_1, s_2, s_3, s_5, s_6 are linked to 4 edges and s_4 to 2 edges).

---

5 Named after the mathematician Leonhard Euler.

2.2.11. *Hamiltonian graph*[6]

A graph is **Hamiltonian** if there is a cycle passing once and only once through all the vertices.

A 1 degree graph cannot be Hamiltonian.

In a graph, if a vertex is 2 degrees, then the two incident edges to this vertex must belong to the Hamiltonian cycle.

Ore's theorem[7]: Let $G=(S, A)$ be a graph of the order $n \geq 3$. If for each vertex x of G, $d(x) \geq n/2$, then the graph G is Hamiltonian.

Dirac's theorem[8]: Let $G=(S, A)$ a graph of the order $n \geq 3$. If for each pair $\{x, y\}$ of non-adjacent vertices, $d(x) + d(y) \geq n/2 + n/2 = n$, then the graph G is Hamiltonian.

2.2.12. *Planar graph*

A graph is **planar** when its edges do not cross.

A topological planar graph is called a **map**. It is said to be connected if its graph is connected. A map divides the plane into a number of **regions**.

The **degree** of a region is the length of its delineating cycle. It is denoted by $d(r)$.

The sum of the degrees of the regions of a connected map is equal to two times the number of edges.

$$\sum_{r=1}^{n} d(r) = 2 \times n_a$$

---

6 Named after the Irish mathematician, physicist and astronomist William Rowan Hamilton, 1805–1865, known for his discovery of quaternions and work on the edge graph of the dodecahedron.
7 Oystein Ore, 1899–1968, Norwegian mathematician known for his work on ring theory, graph theory and Galois connections.
8 Paul Adrien Maurice Dirac, 1902–1984, British physicist and mathematician, founding father of quantum mechanics. He won the Nobel Prize for Physics with Erwin Schrödinger in 1933.

Euler's formula links the number of edges A, the number of vertices S and the number of regions R of a connected map.

$S–A+R=2$

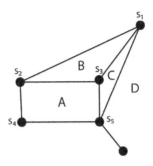

Figure 2.11. *A planar graph*

Figure 2.11 shows a map containing six vertices and eight edges. It divides the plane into four regions: A, B, C and D.

Euler's theorem is confirmed: $6 - 8 + 4 = 2$.

Regions A, B and C are demarcated by the edges of the graph, which is not the case for D which is outside the graph.

The degrees of this graph are $d(A)=4$, $d(B)=3$, $d(C)=3$, $d(E)=5$. If we add up the degrees we get: $4+3+3+6=16$, which confirms what was said previously, $16/2=8$ edges.

Kuratowski's theorem[9] says that a graph is non-planar if and only if it contains a **homeomorphic** subgraph[10] at K_5 (clique on 5 vertices) or $K_{3,3}$ (complete bipartite graph on 3+3 vertices).

---

9 Kazimierz Kuratowski, 1896–1980, Polish mathematician whose work focused on abstract topology and metric space structures among others.
10 An homeomorphism is a continuous bijective application of a topological space in another whose reciprocal bijection is continuous. When two topological spaces are the same but seen differently, they can be said to be homeomorphic.

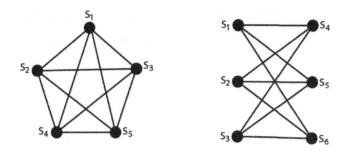

Figure 2.12. *A clique K_5 and a bipartite graph $K_{3,3}$*

2.2.13. *Isthmus*

In a graph, a line is said to be an **isthmus** if its removal increases the number of connected components in the initial graph.

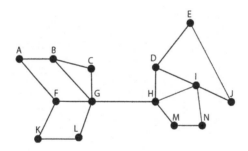

Figure 2.13. *A graph where the edge (G, H) is an isthmus*

In Figure 2.13, if the edge *(G, H)* is removed, there are two components. It can thus be deduced that *(G, H)* is an isthmus. In this graph, no other line is an isthmus.

2.2.14. *Tree and forest*

A **tree** is a connected graph without cycles. A graph without cycles that is not connected is a **forest** (each of its components is a tree).

Figure 2.14. *A tree (left) and a forest (right)*

A tree always has *n-1* edges.

The number of trees it is possible to build with *n* enumerated vertices is n^{n-2}, if $n>1$ (according to Cayley's formula[11]).

Let the graph *G(S, A)* and $x \in S$, it is said that *x* is a **root** of *G* if $\forall y \in S \setminus \{x\}$, there is a path from *x* to *y*.

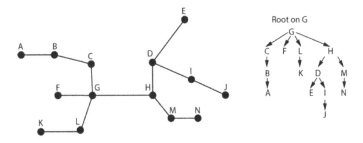

Figure 2.15. *A graph and one of its roots, G*

2.2.15. *Arborescence*

Arborescence is a tree with a distinct vertex called the root. It is often represented following a diagram starting with the root and extending toward the **leaves** located below.

An arborescence can be divided into **levels**. The notion of **rows** also comes into play, which is linked to the vertices.

11 Arthur Cayley, 1821–1895, British mathematician, founder of the Modern British School of Pure Mathematics. He was the first person to introduce multiplication into matrices.

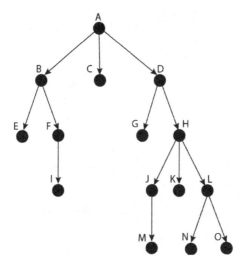

Figure 2.16. *An arborescence*

The **height** of an arborescence is its maximum level or the maximum row that reached one of its vertices.

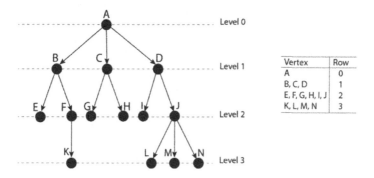

Vertex	Row
A	0
B, C, D	1
E, F, G, H, I, J	2
K, L, M, N	3

Figure 2.17. *An arborescence divided into levels and the rows of its vertices*

2.2.16. *Ordered arborescence*

An **ordered arborescence** is a tree in which all the children for each vertex (node) are completely ordered.

The majority of algebraic expressions can be represented by an ordered arborescence.

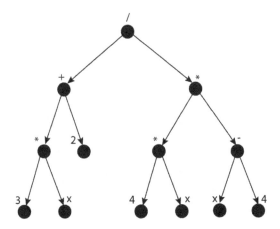

Figure 2.18. *The ordered arborescence of the algebraic expression (3x + 2)/4x(x-4)*

An ordered arborescence is always read from top to bottom and then from left to right. This is very important, as is perfectly illustrated in Figure 2.18: if the graph is read using the leaves *x* and *4* on the right then the result differs from when it is read from *x - 4* or *4 - x*.

There are two methods for traversing an arborescence: **depth-first search** and **breadth-first search**.

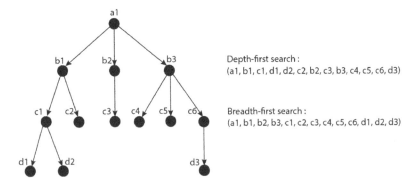

Figure 2.19. *An arborescence and its two searches*

2.3. Directed graph or digraph

When the edges in a graph have a direction, we get a **digraph or oriented graph**. An oriented graph G is formed of two sets, the first S is the set containing the vertices $\{s_1, s_2,..., s_n\}$ and the second C contains the edges $\{c_1, c_2,..., c_n\}$, denoted as $G=(S, C)$.

An edge c is a pair or an ordered couple of vertices. If $c=\{x, y)$, then edge c goes from x to y. The **initial endpoint** of c is y and the **final endpoint** is y.

The **external degree** $d^+(x)$ is denoted as the number of edges with x as the initial endpoint and the **internal degree** $d^-(x)$ is the number of edges with x as a final endpoint, meaning that the degree of a vertex of a diagraph is $d(x)=d^+(x)+d^-(x)$.

A directed graph can be symmetric if the direction of the edges is reversed from the initial graph.

2.3.1. *Path and circuit in a digraph*

A **path** is a sequence of vertices and edges that link one vertex to another.

In a digraph, what we call **distance** is the length of the shortest path linking two vertices.

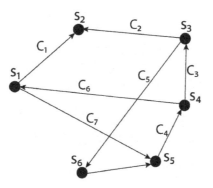

Figure 2.20. *A digraph*

In Figure 2.20, the distance of $d(s4, s1)=1$, $d(s5, s2)=3$, $d(s2, s3)=\infty$ (no path).

A diagraph is said to be **strongly connected** if each vertex can be reached from the other vertices by at least one path.

2.3.2. *Absence of circuit in a digraph*

For applications related particularly to task scheduling, digraphs without circuits are very important. This type of graph is also said to be **acyclic**.

Let $G=(S, C)$ a diagraph. It is without circuit if and only if the set of its vertices has been **topologically sorted**.

Topological sorting is a depth-first search of the graph where a vertex is always visited before its successors, each vertex being given a number denoting the order.

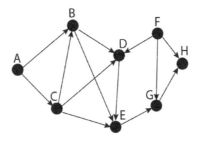

Figure 2.21. *A digraph without circuit*

In Figure 2.21, we consider digraph G=(S, C) with S={A, B, C, D, E, F, G, H} and C={(A, B), (A, C), (B, D), (B, E), (C, B), (C, D), (C, E), (D, F), (D, E), (E, G), (F, G), (F, H), (G, H)}.

An acceptable topological order could be the following sequence of vertices: F, A, C, B, D, E, G, H, meaning that digraph G has no circuit.

A digraph without circuit contains at least a vertex x of the inferior degree or is equal to 0, so $d^{-x}(x) = 0$.

To express the fact that a digraph has no circuit we can also use the notion of **row** $r(x)$ **and level** $n(x)$.

Let G=(S, C) a digraph. Row $r(x)$, with $x \in S$, is the number of edges in the longest path with x as a terminal endpoint. The level $n(x)$, with $x \in S$, is the number of edges in the longest path with x as an initial endpoint.

Each digraph without circuit can be ordered by ascending row or descending level.

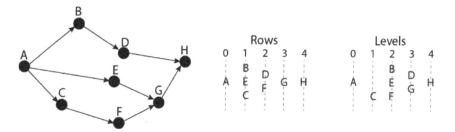

Figure 2.22. *A digraph without circuit with its rows and levels*

2.3.3. Adjacency matrix

A digraph can be represented by an **adjacency matrix**. This is a double entry table with n lines and m columns representing the vertices of the digraph and whose intersections designate a vertex. This matrix is always square and it always has 0 on its diagonal unless it is a loop. It is not symmetric.

On the right, Figure 2.23 shows the adjacency matrix representing the graph. When an edge joins two edges the value in the matrix is 1.

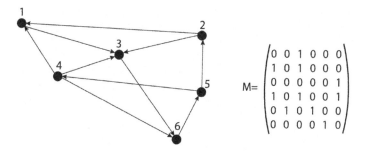

$$M = \begin{pmatrix} 0 & 0 & 1 & 0 & 0 & 0 \\ 1 & 0 & 1 & 0 & 0 & 0 \\ 0 & 0 & 0 & 0 & 0 & 1 \\ 1 & 0 & 1 & 0 & 0 & 1 \\ 0 & 1 & 0 & 1 & 0 & 0 \\ 0 & 0 & 0 & 0 & 1 & 0 \end{pmatrix}$$

Figure 2.23. *An example of a graph and its adjacency matrix*

There are other possible uses for the adjacency matrix, which has very interesting properties. These uses will be described in the following chapters of this book.

2.3.4. *Valued graph matrix*

As for the adjacency matrix, a valued graph can be represented by a square matrix. Each coefficient corresponds to the value (weight, cost) represented by an edge.

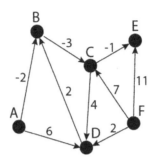

Figure 2.24. *A valued digraph*

If $G = (S, C, v)$ is the graph in Figure 2.24, the **valuation matrix** of G can be defined as the square matrix $M=m(i, j)$ with a size $n \times n$ respecting:

$$M_{ij} = v(i, j) \ if \ (i, j) \in C \ \text{otherwise} \ \infty$$

We get the associated matrix:

$$\begin{bmatrix} \infty & -2 & \infty & 6 & \infty & \infty \\ \infty & \infty & -3 & \infty & \infty & \infty \\ \infty & \infty & \infty & 4 & -1 & \infty \\ \infty & 2 & \infty & \infty & \infty & \infty \\ \infty & \infty & \infty & \infty & \infty & \infty \\ \infty & \infty & 7 & 2 & 11 & \infty \end{bmatrix}$$

2.4. Graphs for logistics

Throughout this book you will be presented with countless graphs of all types depending on the sort of problem they deal with.

Logistics cannot do without graphs and uses them to represent functions or simple processes whatever the chosen level within the enterprise.

This type of representation goes well beyond this and is not only the prerogative of logistics and operations research. It is fascinating to see how a simple graph can replace an explanation that it is often difficult to formulate with words.

As Napoleon Bonaparte said, "*A good sketch is better than a long speech*".

3

Optimal Paths

3.1. Basic concepts

When two points in a network need to be joined, whether it is a road, river, IT, electric or other type of network such as to optimize a cost, journey time or distance, optimal paths are often used.

Within planning, scheduling and flows in graphs, they are necessary for solving certain types of – often underlying – problems.

In the domain of videogames, the virtual opponents of the player must move around often complex environments while maintaining maximum gameplay and guaranteeing an acceptable level of difficulty. The developer of these systems must enable these entities to move around quickly and reliably, enabling them to fight effectively against the player: optimal paths work wonders.

Several resolution methods are possible:

– **Dijkstra's's**[1] **algorithm;**

– **Flyod–Warshall's**[2] **algorithm;**

– **Bellman-Ford's**[3] **algorithm.**

---

1 Edsger Wybe Dijkstra was a Dutch computer scientist (1930–2002). This algorithm uses the principle of dynamic programming (see Chapter 4).
2 Also called the Roy–Flyod algorithm.
3 Like Dijkstra's algorithm, this algorithm is also based on the principle of dynamic programming (see Chapter 4).

As their **complexities** differ, these algorithms do not attain the same level of effectiveness. They have complexities[4] which are $O(n^2)$, $O(n^3)$ and $O(n^4)$, respectively.

3.2. Dijkstra's algorithm

Dijkstra's algorithm may or may not be applied to a connected directed graph (digraph). The weight of each edge must be positive or null.

It enables the shortest paths from one vertex to all the others to be found.

Each vertex will receive a couple of values *(d(s), p(s))* where *d(s)* represents the distance, cost, duration, weight, depending on the type of problem, and *p(s)* the immediate predecessor vertex.

It is based on a progressive algorithm and follows these steps:

1) Initialization: The couple *(0, -)* is attributed to the starting vertex, which is marked. The other vertices benefit from the couple *(∞, -)*. As this vertex is marked, it will no longer be the object of other calculations. It is said to be **saturated**.

2) The vertices adjacent to the starting vertex are attributed a couple of values *(d(s), p(s))* where *d(s)* is the value of the edge linking them and *p(s)* the immediate predecessor vertex. The non-adjacent vertices maintain their previous value.

3) Among the couples, the couple with the smallest distance is sought out *d(s)*, it is marked and saturated. In the event of duplication, one is chosen by default.

4) The vertices adjacent to this marked vertex are sought out and are attributed a couple *(d(s), p(s))*. Distance *d(s)* is the sum of the distance of the vertex that has just been marked and the value of the edge connecting it, unless it is lower or equal to the distance that exists at the previous iteration, otherwise the latter is noted. The value of *p(s)* is its immediate predecessor vertex. With regards to non-adjacent vertices, the value obtained at the previous iteration is noted.

---

4 The complexity of an algorithm is the number of elementary operations it must perform. It is denoted by O and is based on the size of the data *n*.

5) Steps 3 and 4 are repeated until all the vertices have been consumed or until the end vertex.

3.2.1. *An example of calculating minimal paths*

Let us consider the graph below which represents a road network made up of 8 towns from A to H. The edges represent the possible communication routes between each of these towns and show the distance in kilometers. The starting vertex is town A and the end vertex is town H.

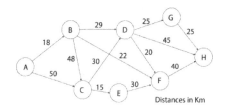

Figure 3.1. *An example of a road network*

Table 3.1 shows the different iterations (lines) of the algorithm for each of the vertex-distance, immediate predecessor couples *(d(s), p(s))*. The marked couples and saturated columns are gray.

Stage/Distance, Predecessor	d(A),p(A)	d(B),p(B)	d(C),p(C)	d(D),p(D)	d(E),p(E)	d(F),p(F)	d(G),p(G)	d(H),p(H)	Marked vertex
Initialization	(0, -)	(∞, -)	(∞, -)	(∞, -)	(∞, -)	(∞, -)	(∞, -)	(∞, -)	A
1		(18, A)	(50, A)	(∞, -)	(∞, -)	(∞, -)	(∞, -)	(∞, -)	A,B
2			(50, A)	(47, B)	(∞, -)	(40, B)	(∞, -)	(∞, -)	A,B,F
3			(50, A)	(47, B)	(70, F)		(65, D)	(80, F)	A,B,F,D
4			(50, A)		(70, F)		(65, D)	(80, D)	A,B,F,D,C
5					(65, C)		(65, D)	(80, D)	A,B,F,D,C,E
6							(65, D)	(80, D)	A,B,F,D,C,E, G
7								(80, D)	A,B,F,D,C,E, G,H

Table 3.1. *The iterations of the algorithm*

3.2.2. *Interpreting the results of the calculations*

To find out the minimal distance between town A and another town, read the value *d(s)* of the marked couple in the corresponding column. For instance, the minimal distances from A to H and A to E are 80 and 65 km, respectively.

To determine the route, go to the town of arrival column, read the value of *p(s)*, the predecessor of the marked couple, and note it. Look in the column containing the noted *p(s)* and then read once again the value of *p(s)* of the marked couple, and so on and so forth until the first vertex is reached, in this case A. For example, the route of H toward A will be: H, D, B, A.

– H, end vertex,

– D, value of *p(s)* of the marked couple in column H,

– B, value of *p(s)* of the marked couple in column D,

– A, value of *p(s)* of the marked couple in column B.

3.3. Flyod–Warshall's algorithm

In this algorithm, the graph must be directed, valued and not have a negative weight cycle. The edges can bring negative values (distances, weight, cost, etc.).

It enables the shortest paths for all pairs of vertices to be found.

It is based on the creation of two matrices, the first for distances M_d and the second for predecessors M_p on which a set of iterations will be performed after an initialization stage where the matrices are constructed.

3.3.1. *Creating the starting matrices (initialization of the algorithm)*

The distance matrices M_d will be filled by the distances *d(s)* brought by the directed edges which may exist between a couple of vertices *c(i, j)* in the following conditions:

$$\exists\, c(i,j) \Rightarrow M_d(i,j) = d(s)$$

$$\nexists c(i,j) \Longrightarrow M_d(i,j) = \infty$$

$$i = j \Longrightarrow M_d(i,j) = 0$$

COMMENT 3.1.– The existence of an edge is considered according to its direction. By applying the previous conditions, the diagonal of the matrix will only contain null values.

The values contained in the predecessor matrix M_p are defined according to the following conditions:

$$\exists\, c(i,j) \Longrightarrow M_p(i,j) = i$$

$$\nexists c(i,j) \Longrightarrow M_p(i,j) = \emptyset$$

$$i = j \Longrightarrow M_p(i,j) = i$$

COMMENT 3.2.– By applying the previous conditions, the diagonal of the matrix links the list of the vertices.

3.3.2. Filling the matrices for the following iterations

To complete matrices M_d and M_p in the following iterations, the number of vertices n_s in the graph is considered and we apply the algorithm composed of three nested loops: k (number of iterations), i (for the starting vertices of an edge) and j (for the end vertices of an edge).

```
for k = 1 to nₛ do
      for i = 1 to nₛ do
            for j = 1 to nₛ do
                  Md(i, j)ᵏ = min{ Md(i, j)ᵏ⁻¹, Md(i, k)ᵏ⁻¹ + Md(k, j)ᵏ⁻¹}
                  if Md(i, j)ᵏ = Md(i, k)ᵏ⁻¹ + Md(k, j)ᵏ⁻¹ then
                        Mp(i , j)ᵏ = Mp(k , j)ᵏ⁻¹
                  otherwise
                        Mp(i , j)ᵏ = Mp(i , j)ᵏ⁻¹
                  end if
            end for
      end for
end for
```

3.3.3. *An example of calculating minimal paths*

Let us consider the road network in Figure 3.2, showing the possible links between 5 towns represented by the vertices 1, 2, 3, 4 and 5. The distances are expressed in kilometers.

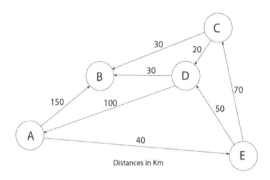

Figure 3.2. *The road network used for our example*

COMMENT 3.3.– The towns in our example are represented by vertices enumerated from 1 to 5, although it would seem more logical to attribute them letters. However, for convenience and the comprehension of the steps in the algorithm presented below, figures are better suited.

The steps in the algorithm follow the iterations described below. The table on the left represents the distance matrix and the table on the right the predecessor matrix. The modifications occurring along the sequence are colored gray.

Initialization

Vertex	1	2	3	4	5		1	2	3	4	5
1	0	150	∞	∞	40		1	1	-	-	1
2	∞	0	∞	∞	∞		-	2	-	-	-
3	∞	30	0	20	∞		-	3	3	3	-
4	100	30	∞	0	∞		4	4	-	4	-
5	∞	∞	70	50	0		-	-	5	5	5

Table 3.2. *Initialization*

Iteration no. 1

Vertex	1	2	3	4	5
1	0	150	∞	∞	40
2	∞	0	∞	∞	∞
3	∞	30	0	20	∞
4	100	30	∞	0	140*
5	∞	∞	70	50	0

1	2	3	4	5
1	1	-	-	1
-	2	-	-	-
-	3	3	3	-
4	4	-	4	1*
-	-	5	5	5

Table 3.3. *Iteration no. 1*

Details for calculating the edge (4, 5):

Let $k=1$, $i=4$ and $j=5$

We have:

$$M_d(4, 5)^1 = min\{M_d(4, 5)^{1-1}, M_d(4, 1)^{1-1} + M_d(1, 5)^{1-1}\}$$

$$M_d(4, 5)^1 = min\{\infty, 100+40\} = 140$$

If $M_d(4, 5)^1 = M_d(4, 1)^{1-1} + M_d(1, 5)^{1-1}$ then $M_p(4, 5)^1 = M_p(1, 5)^{1-1} = 1$

Iteration no. 2

Vertex	1	2	3	4	5
1	0	150	∞	∞	40
2	∞	0	∞	∞	∞
3	∞	30	0	20	∞
4	100	30	∞	0	140
5	∞	∞	70	50	0

1	2	3	4	5
1	1	-	-	1
-	2	-	-	-
-	3	3	3	-
4	4	-	4	1
-	-	5	5	5

Table 3.4. *Iteration no. 2*

Iteration no. 3

Vertex	1	2	3	4	5
1	0	150	∞	∞	40
2	∞	0	∞	∞	∞
3	∞	30	0	20	∞
4	100	30	∞	0	140
5	∞	10	70	50	0

1	2	3	4	5
1	1	-	-	A
-	2	-	-	-
-	3	3	3	-
4	4	-	4	1
5	3	5	5	5

Table 3.5. *Iteration no. 3*

Iteration no. 4

Vertex	1	2	3	4	5
1	0	150	∞	∞	40
2	∞	0	∞	∞	∞
3	120	30	0	20	160
4	100	30	∞	0	140
5	150	80	70	50	0

1	2	3	4	5
1	1	-	-	1
-	2	-	-	-
4	3	3	3	-
4	4	-	4	1
4	4	5	5	5

Table 3.6. *Iteration no. 4*

Iteration no. 5

Vertex	1	2	3	4	5
1	0	120	110	90	40
2	∞	0	∞	∞	∞
3	120	30	0	20	160
4	100	30	210	0	140
5	150	80	70	50	0

1	2	3	4	5
1	4	5	5	1
-	2	-	-	-
4	3	3	3	-
4	4	5	4	1
4	4	5	5	5

Table 3.7. *Iteration no. 5*

3.3.4. *Interpreting the results*

To determine the minimum distance between 2 vertices in the graph, consult the number at the line–column intersection of the distance matrix M_d in the last iteration.

For instance:

– to go from town 1 to town 2, we find 120 km;

– to go from town 5 to town 2, we find 80 km;

– to go from town 4 to town 3, we find 210 km.

To determine the path, retrace the route step-by-step from the starting vertex to the end vertex.

Let us consider the route from 1 to 2:

– we go to the predecessor matrix M_p, we consult the cell at the intersection of 1 and 2: we find 4;

– in this same matrix, we then go to the intersection of 1 and 4: we find 5;

– we continue by going to the intersection of 1 and 5: we find 1. We have therefore reached the starting vertex;

– the path is therefore 1, 5, 4 and 2 (i.e. the sum of the distances: 40 + 50 + 30 = 120 km).

Let us consider the route from 5 to 2:

– we go to M_p and at the intersection of 5 and 2 we read the value 4;

– we go to the intersection of 5 and 4: we find 5. The starting vertex is reached;

– the path is 5, 4 and 2 (i.e. 50 + 30 = 80 km).

Let us consider the route from 4 to 3:

– we go to M_p and at the intersection of 4 and 3 we read the value 5;

– we go to the intersection of 4 and 5: we find 1;

– we go to the intersection of 4 and 1: we find 4. The starting vertex is reached;

– the path is 4, 1, 5 and 3 (i.e. $100 + 40 + 70 = 210$ km).

3.4. Bellman–Ford's algorithm

This algorithm also solves the problems of finding the shortest paths in a directed graph from one origin. The edges can bring negative values.

It finds the set of the shortest paths from one vertex to all other vertices.

The steps in the algorithm are based on an operation called **relaxation**, which relays a true or false value.

The number of iterations is equal to $n-1$ vertices in the graph, minus one, in the event that the algorithm converges, i.e., there is no negative circuit (see section 3.5).

3.4.1. Initialization

We construct a table of vertices which collects together the distances $d(s)$ and the predecessors $p(s)$ of each of the vertices in the graph:

– we choose the starting vertex and give it a distance of null (of this vertex toward itself), $d(s)=0$. As the latter does not have a predecessor, $p(s)=\varnothing$;

– the other vertices are initialized with a distance $d(s)=\infty$ and a predecessor $p(s)=\varnothing$.

Vertex s	s_1	...	s_n
$d(s)$	0	∞	∞
$p(s)$	$\varnothing$	$\varnothing$	$\varnothing$

Table 3.8. *Initialization of the vertices*

3.4.2. *The next iterations with relaxation*

The next iterations are based on two nested loops, where n is the number of vertices in the graph and $c(i, j)$, a couple representing an valued edge:

```
for s = 1 to n-1 do
        for each c(i, j)
                if d(i)ˢ⁻¹ + c(i, j) < d(j)ˢ⁻¹ then
                        {relaxation operation}
                        d(j)ˢ = d(i)ˢ⁻¹ + c(i, j)
                        p(j) = i
                end if
        end for
end for
```

A table can collect together the results of the test and the relaxation operation for all the edges in the graph at each iteration of the loop s. The calculated values can then be added to the table of the vertices created during initialization, which will consequently grow.

3.4.3. *An example of calculation*

Let us return to the graph in example section 3.3.3. The towns are now represented by letters A, B, C, D and E, the distances remain the same.

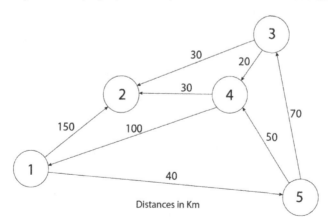

Figure 3.3. *The graph of our example with 5 towns A, B, C, D, E and F*

Table of the vertices at initialization

Vertex s	A	B	C	D	E
$d(s)$	0	∞	∞	∞	∞
$p(s)$	∅	∅	∅	∅	∅

Table 3.9. *Initialization of the vertices in the example*

Table of relaxation at iteration no. 1

Edge	Distance	Test	$d(j)$	$p(j)$
AB*	150	true	150	A
AE	40	true	40	A
CB	30	false		
CD	20	false		
DA	100	false		
DB	30	false		
EC	70	true	110	E
ED	50	true	90	E

Table 3.10. *Table of relaxation for iteration no. 1*

Table of the vertices at iteration no. 1

Vertex s	A	B*	C	D	E
$d(s)$	0	150	110	90	40
$p(s)$	∅	A	E	E	A

Table 3.11. *Table of the vertices, iteration no. 1*

Details for calculating the first line, (edge AB):

Let $d(A)^0=0$, $d(B)^0=∞$ and $c(A,B)=150$

We have: if $0+150<∞$ *then* relaxation which means that d(B)=150 and p(B)=A.

Table of relaxation at iteration no. 2

Edge	Distance	Test	$d(j)$	$p(j)$
AB	150	false		
AE	40	false		
CB	30	true	140	C
CD	20	false		
DA	100	false		
DB	30	true	120	D
EC	70	false		
ED	50	false		

Table 3.12. *Table of relaxation for iteration no. 2*

Table of the vertices at iteration no. 2

Vertex s	A	B	C	D	E
$d(s)$	0	120	110	90	40
$p(s)$	$\varnothing$	D	E	E	A

Table 3.13. *Table of the vertices, iteration no. 2*

Table of relaxation at iteration no. 3

Edge	Distance	Test	$d(j)$	$p(j)$
AB	150	false		
AE	40	false		
CB	30	false		
CD	20	false		
DA	100	false		
DB	30	false		
EC	70	false		
ED	50	false		

Table 3.14. *Table of relaxation for iteration no. 3*

Table of the vertices at iteration no. 3

Vertex s	A	B	C	D	E
$d(s)$	0	120	110	90	40
$p(s)$	$\varnothing$	D	E	E	A

Table 3.15. *Table of the vertices, iteration no. 3*

COMMENT 3.4.– It is not necessary to continue until the final iteration as responses to the test were false for each of the edges.

3.4.4. *Interpreting the results*

To find out the minimum distance between the starting vertex A and another vertex, consult the final table of the vertices calculated. For example, the distance from vertex A to vertex C is equal to the value of column C, line $d(s)$, i.e. 110 km.

To find the path associated to this value, look in column C, the line $p(s)$ of the predecessors, i.e. the vertex E. Then go to column E and in the same way look at the value of $p(s)$, i.e. A.

The path from A to C is therefore A, E, C.

By following this approach, all the minimum distances and their paths can be found for each of the routes beginning on A.

3.5. Bellman–Ford's algorithm with a negative circuit

As mentioned in section 3.3, Bellman–Ford's algorithm can detect a negative circuit of cost (weight) in a directed graph.

This type of circuit is often useful for resolving financial problems, which require decision support or arbitrage opportunities. Numerous flows problems also call upon this particularity.

The method used is identical to the method presented above.

During the final iteration one or more potential relaxation calculations are detected thus showing that a negative circuit is present in the graph. The algorithm will never converge.

3.5.1. *Example*

Let us consider the graph below:

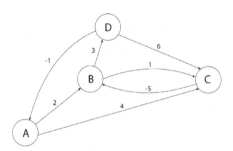

Figure 3.4. *An example of a graph with a negative circuit*

Table of the vertices at initialization

Vertex s	A	B	C	D
$d(s)$	0	∞	∞	∞
$p(s)$	∅	∅	∅	∅

Table 3.16. *Table of vertices, initialization*

Table of relaxation at iteration no. 1

Edge	Distance	Test	$d(j)$	$p(j)$
AB	2	true	0+2=2	A
AC	4	true	0+4=4	A
BC	1	true	2+1=3	B
BD	3	true	2+3=5	B
CB	-5	true	3+-5=-2	C
DA	-1	false		
DC	6	false		

Table 3.17. *Table of relaxation for iteration no. 1*

Table of the vertices at iteration no. 1

Vertex s	A	B	C	D
$d(s)$	0	-2	3	5
$p(s)$	∅	C	B	B

Table 3.18. *Table of the vertices, iteration no. 1*

Table of relaxation at iteration no. 2

Edge	Distance	Test	$d(j)$	$p(j)$
AB	2	false		
AC	4	false		
BC	1	true	-2+1=-1	B
BD	3	true	-2+3=1	B
CB	-5	true	-1+-5=-6	C
DA	-1	false		
DC	6	false		

Table 3.19. *Table of relaxation for iteration no. 2*

Table of the vertices at iteration no. 2

Vertex s	A	B	C	D
$d(s)$	0	-6	-1	1
$p(s)$	∅	C	B	B

Table 3.20. *Table of the vertices, iteration no. 2*

Table of relaxation at iteration no. 3

Edge	Distance	Test	$d(j)$	$p(j)$
AB	2	false		
AC	4	false		
BC	1	true	-5	B
BD	3	true	-3	B
CB	-5	true	-10	C
DA	-1	true	-4	D
DC	6	false		

Table 3.21. *Table of relaxation for iteration no. 3*

Table of vertices at iteration no. 3

Vertex s	A	B	C	D
$d(s)$	-4	-10	-5	-3
$p(s)$	D	C	B	B

Table 3.22. *Table of vertices, iteration no. 3*

During this third iteration, the algorithm should have converged given that, according to Bellman–Ford, it should be limited to a number of iterations equal to n-1 vertices. In this case, all the tests should be *false*, which does not happen, thus indicating that in the graph in question there is at least one negative circuit.

Let us continue the sequence by going onto iteration no. 4

Table of relaxation at iteration no. 4

Edge	Distance	Test	$d(j)$	$p(j)$
AB	2	false		
AC	4	false		
BC	1	true	-9	B
BD	3	true	-7	B
CB	-5	true	-14	C
DA	-1	true	-8	D
DC	6	false		

Table 3.23. *Table of relaxation for iteration no. 4*

Table of vertices at iteration no. 4

Vertex s	A	B	C	D
$d(s)$	-8	-14	-9	-7
$p(s)$	D	C	B	B

Table 3.24. *Table of vertices, iteration no. 4*

Once again, not all the tests are *false*. If we were to continue, the negative values would become increasingly higher, continuing infinitely to decrease.

The presence of a negative circuit can be checked in our example. We find out the path between A and D, by applying the same technique described in section 3.4.4 to get: ...B, C, B, D. We note the presence of an infinite loop B, C, B.

3.6. Exercises

3.6.1. *Exercise 1: optimizing journey time*

The graph below represents a metro network. The seven stations A, B, C, D, E, F and G are connected by one or a number of lines whose journey time is noted on each of the edges (for the purpose of this exercise, we will not take into account the waiting times for connections at each station).

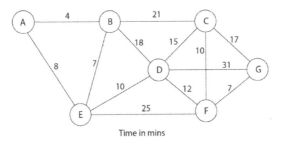

Figure 3.5. *A graph representing a metro network*

Questions:

1) Using Dijkstra's algorithm, determine the shortest time to get from station A to station G.

2) What is the shortest path?

3.6.2. *Exercise 2: a directed graph with negative cost edge*

Let us consider the graph below:

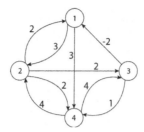

Figure 3.6. *A graph with a negative cost edge*

Questions:

1) Determine the minimum path and weight between vertices 4 and 1 by applying Flyod–Warshall's algorithm.

2) In the same way, determine the minimum path and weight between vertices 2 and 1.

3.6.3. *Exercise 3: routing data packets*

Two servers are connected by the network below, constituted of 6 nodes A, B, C, D, E and F. The routers in this network use the routing protocol RIP[5] which is based on Bellman–Ford's algorithm. The minimum number of hops is indicated on each of the edges.

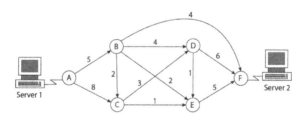

Figure 3.7. *The network connecting the 2 servers*

Question:

1) Determine the optimum path and its cost for establishing a link between the two servers.

5 Routing Information Protocol is "distance vector" type IP routing protocol i.e. each router communicates to its neighbors the number of jumps or "hops" (distance) separating them.

3.6.4. *Solution for exercise 1*

Question 1:

The different iterations of the algorithm are collected together in Table 3.25:

Vertex	A.	B	C	D	E	F	G
Initialization	(0, -)	(∞, -)	(∞, -)	(∞, -)	(∞, -)	(∞, -)	(∞, -)
1	•	(4, A)	(∞, -)	(∞, -)	(8, A)	(∞, -)	(∞, -)
2	•	•	(25, B)	(22, B)	(8, A)	(∞, -)	(∞, -)
3	•	•	(25, B)	(18, E)	•	(33, E)	(∞, -)
4	•	•	(25, B)	•	•	(30, D)	(49, D)
5	•	•	•	•	•	(30, D)	(40, F)
6	•	•	•	•	•	•	(37, F)

Table 3.25. *The iterations of the algorithm*

To get from A to G, that the journey will last 37 min.

Question 2:

The shortest path from A to G is: A, E, D, F, G.

3.6.5. *Solution for exercise 2*

Question 1:

The different iterations of the algorithm are displayed in Tables 3.26 to 3.30.

Initialization

Vertex	1	2	3	4
1	0	3	∞	3
2	2	0	2	2
3	-2	∞	0	1
4	∞	4	4	0

1	2	3	4
1	1	-	1
2	2	2	2
-	3	3	3
-	4	4	4

Table 3.26. *Initialization*

Iteration no. 1

Vertex	1	2	3	4
1	0	3	∞	3
2	2	0	2	2
3	-2	1	0	1
4	∞	4	4	0

1	2	3	4
1	1	-	1
2	2	2	2
3	1	3	3
-	4	4	4

Table 3.27. *Iteration no. 1*

Iteration no. 2

Vertex	1	2	3	4
1	0	3	5	3
2	2	0	2	2
3	-2	1	0	1
4	6	4	4	0

1	2	3	4
1	1	2	1
2	2	2	2
3	1	3	3
2	4	4	4

Table 3.28. *Iteration no. 2*

Iteration no. 3

Vertex	1	2	3	4
1	0	3	5	3
2	0	0	2	2
3	-2	1	0	1
4	2	4	4	0

1	2	3	4
1	1	2	1
3	2	2	2
3	1	3	3
3	4	4	4

Table 3.29. *Iteration no. 3*

Iteration no. 4

Vertex	1	2	3	4
1	0	3	5	3
2	0	0	2	2
3	-2	1	0	1
4	2	4	4	0

1	2	3	4
1	1	2	1
3	2	2	2
3	1	3	3
3	4	4	4

Table 3.30. *Iteration no. 4*

The path between 4 and 1 is: 4, 3, 1 and its weight is 2.

Question 2:

The path between 2 and 1 is: 2, 3, 1 and its weight is 0.

3.6.6. *Solution for exercise 3*

Table of the vertices at initialization

Vertex s	A	B	C	D	E	F
$d(s)$	0	∞	∞	∞	∞	∞
$p(s)$	$\varnothing$	$\varnothing$	$\varnothing$	$\varnothing$	$\varnothing$	$\varnothing$

Table 3.31. *Initialization*

Table of relaxation at iteration no. 1

Edge	Distance	Test	$d(j)$	$p(j)$
AB	5	true	5	A
AC	8	true	8	A
BC	2	true	7	B
BD	4	true	9	B
BE	2	true	7	B
BF	4	true	9	B
CD	3	false		
CE	1	false		
DE	1	false		
DF	6	false		
EF	5	false		

Table 3.32. *Table of relaxation, iteration no. 1*

Table of the vertices at iteration no. 1

Vertex s	A	B	C	D	E	F
$d(s)$	0	5	7	9	7	9
$p(s)$	∅	A	B	B	B	B

Table 3.33. *Table of the vertices, iteration no. 1*

Table of relaxation at iteration no. 2

Edge	Distance	Test	$d(j)$	$p(j)$
AB	5	false		
AC	8	false		
BC	2	false		
BD	4	false		
BE	2	false		
BF	4	false		
CD	3	false		
CE	1	false		
DE	1	false		
DF	6	false		
EF	5	false		

Table 3.34. *Table of relaxation, iteration no. 2*

Table of the vertices at iteration no. 2

Vertex s	A	B	C	D	E	F
$d(s)$	0	5	7	9	7	9
$p(s)$	∅	A	B	B	B	B

Table 3.35. *Table of the vertices, iteration no. 2*

As the responses to the test were false for each of the edges, there is no need to continue after iteration no. 2.

The path of server no. 1 to server no. 2 is A, B, F and its cost is 9.

4

Dynamic Programming

4.1. The principles of dynamic programming

Dynamic programming is an optimization method based on the principle of optimality defined by Bellman[1] in the 1950s: *"An optimal policy has the property that whatever the initial state and initial decision are, the remaining decisions must constitute an optimal policy with regard to the state resulting from the first decision."*

It can be summarized simply as follows: "every optimal policy consists only of optimal sub policies."

It is a very powerful technique, but its application framework is limited. Nevertheless, numerous variants exist to best meet the different problems encountered.

This method is a variant of the "divide and conquer" method given that a solution to a problem depends on the previous solutions obtained from subproblems. The main and major difference between these two methods relates to the superimposition of subproblems in dynamic programming. A subproblem can be used to solve a number of different subproblems. In the "divide and conquer" approach, subproblems are entirely independent and can be solved separately. Moreover, recursion is used, unlike in dynamic

---

1 Richard Bellman, 1920–1984 was an American mathematician. He is considered to be the inventor of dynamic programming.

programming where a combination of small subproblems is used to obtain increasingly larger subproblems.

To sum up, it can be said that the "divide and conquer" method works by following a top-down approach whereas dynamic programming follows a bottom-up approach.

How these two methods function can be illustrated and compared in two arborescent graphs.

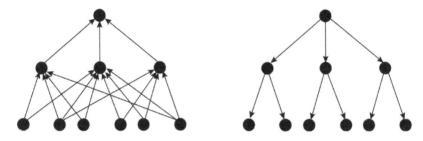

Figure 4.1. *The methods: dynamic programming (left) and divide and conquer (right)*

Problems concerning manufacturing management and regulation, stock management, investment strategy, macro planning, training, game theory, computer theory, systems control and so on result in decision-making that is regular and based on sequential processes which are perfectly in line with dynamic programming techniques.

4.2. Formulating the problem

A dynamic programming algorithm generally consists of a number of phases that link together to arrive at the optimal solution: characterization, definition, calculation and construction.

To study the elements of dynamic programming, we are going to go over a number of simple examples for which we will formulate the problem and then apply the principle of optimality such as to arrive at a solution.

4.2.1. *Example 1 – the pyramid of numbers*

Let us consider the pyramid of numbers below:

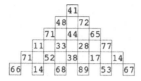

Figure 4.2. *The initial pyramid of numbers to traverse*

Find the maximum sum of the numbers by traversing the pyramid from top to bottom. Only one number per line is authorized and the boxes containing the numbers must share a summit.

If N represents the number of boxes, the total number of possible paths is equal to 2N-1, in this case N=21:

$2 \times 21 = 41$ paths

First, a simple, evident and elegant solution can be found. All that is required is to add all the possible variations of values while bearing in mind that each number has in the row above one or two ascendants depending on whether it is at the end of the line or not. The summit of the pyramid obviously does not have an ascendant.

Here is the tree of all the possibilities:

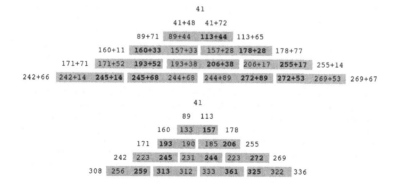

Figure 4.3. *The tree of the possibilities with the possible calculations (above) and their results (below). The highest totals are indicated in bold*

When two ascendants exist, only the highest total is retained. Continuing in this way, we arrive at the pyramid shown in Figure 4.4.

				41				
			89		113			
		160		157		178		
	171		193		206		255	
242		245		244		272		269
308	259		313		361		325	336

Figure 4.4. *The pyramid obtained by retaining only the maximum totals in each line. The numbers in bold are the maximal values in each line*

The maximal total is 361. It can be obtained by adding these figures from the initial pyramid: $41 + 72 + 65 + 77 + 17 + 89 = 361$.

The fact of applying the strategy described above, making use – throughout the search of the pyramid – of the results of the intermediary totals (subproblems) and only retaining the highest number to move forward, towards the next line, relates to the principle of dynamic programming.

4.2.2. Example 2 – the Fibonacci[2] sequence

The Fibonacci sequence is a sequence of integers in which each number is the sum of the two previous numbers. It starts with the numbers 0 and 1.

0+1=1, 1+1=2, 2+1=3, 3+2=5, 5+3=8, 8+5=13, 13+8=21, 21+13=34,...

To calculate the sequence, a simple algorithm can be defined by calling upon recursion[3].

Fibonacci procedure *(n)*
      if $n \leq 1$ then
            note *n*
      otherwise
            fibonacci(n – 1) + fibonacci(n – 2)
      end if
end of procedure

---

2 Leonardo Fibonacci, 1175–1250, was an Italian mathematician who wrote numerous works. The famous "Fibonacci sequence" stems from a problem described in the book "Liber Abaci", published in 1202.
3 A recursive algorithm is one that calls itself.

This algorithm has a strong complexity due to the multiplicity of identical calculations that are realized a high number of times.

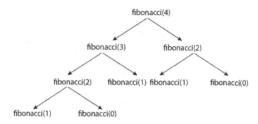

Figure 4.5. *Multiplicity of calculations in the execution of the algorithm*

The performance of this algorithm can be improved if we memorize the intermediary results such that they can be reused when needed without having to recalculate.

Implementing this technique is similar to dynamic programming.

```
Fibonacci procedure (n, table[ ])
        if n ≤ 1 then
                note n
        end ifi
        if table[n] does not exist then
                table[n] = fibonacci(n – 1) + fibonacci(n – 2)
        end if
        note table[n]
end of procedure
```

This technique drastically reduces complexity.

We can progress further by removing recursion and filling the *table[n]* space by space. However, this requires processing to be reversed: we start with the smallest values.

```
Fibonacci procedure (n)
        size table[n] with n+1 spaces
        table [0]=0
        table[1]=1
        for i from 2 to n-1 do
                table [i] = table[i - 1] + table[i -2]
        end for
        note table[n]
end of procedure
```

In this final algorithm, we are using the concepts of dynamic programming perfectly.

4.2.3. *Example 3 – the knapsack*

The knapsack problem (**KP**) is a very well-known problem in computer science. It occurs in many situations be they in industry, finance, the applied sciences or in real life.

It was very widely studied over the course of the 20th Century and today we continue to work on its resolution methods and applications.

Dynamic programming is well adapted to this problem for which we can conceive an optimal solution at *n* variables from a subproblem at *n-1* variables.

4.2.3.1. *Presenting the problem*

A knapsack has a maximum load capacity. We have different objects of different weights and values to fill it with. How can we fill up this knapsack such that it holds a maximal total weight and total value? These two elements must be defined as well as the list of objects that will fill up the knapsack.

It should be noted that dynamic programming is not the only method to find a solution. Other methods can be used such as **genetic algorithms, greedy algorithms** or algorithms based on **BB (branch and bound)**.

4.2.3.2. *Resolution algorithm*

For this algorithm, we will use the following variables:

– C, maximum capacity of the knapsack;

– i, increment of the line of the table;

– j, increment of the column of the table;

– n, maximum number of objects in the knapsack;

– *table[i, j]*, matrix containing the maximum benefits for the objects with as input on the abscissa, the weight of the object, and on the ordinate, the objects from A to n;

– *p[ ]*, vector (table of dimension 1) containing the weight of each of the objects;

– *v[ ]*, vector containing the value of each of the objects.

The steps of the algorithm are presented below. First, we will fill the matrix which we will then read starting from the last box to determine the total value, the maximum weight and the list of the objects.

Phase 1: Filling the matrix

```
For j from 0 to C do
        table[0, j] = 0
end for
for i from 1 to n do
     for j from 0 to C do
          if j ≥ p[i] then
                  table[i, j] = max(table[i - 1, j] ; table[i - 1, j – p[i]] + v[i])
          otherwise
                  table[i, j] = table[i – 1, j]
          end if
     end if
end if
```

The final box of the *table* matrix contains the total maximum value of all the objects that we put in the knapsack.

Phase 2a: Reading the maximum weight

```
for k from C to 0 do
     if table[i, j] = table[i, j-1)]
end for
```

At the end of the loop *k, j* is equal to the total maximum weight (index of the column) that the knapsack can contain, which is always less or equal to *C*.

Phase 2b: Obtaining the list of objects

```
as long as i>0 and j<0 do
     if table[i, j] ≠ table[i-1, j] then
          j=j-p(i)
          i=i-1
```

otherwise
    $i=i-1$
end it
end as long as

In the event the test is positive, i is equal to the number of the object (index of the line) to be added to the knapsack to obtain the total maximum weight.

4.2.3.3. *An example of an application*

Let us consider a knapsack with a maximum capacity of 14 kg that we can fill with six objects A, B, C, D, E and F whose weights and values are laid out in Table 4.1.

Object	A	B	C	D	E	F
Value	7	8	14	5	10	15
Weight	2	1	5	2	4	3

Table 4.1. *Objects, values and weight*

What maximum value of objects can we put in the knapsack without exceeding its capacity?

Phase 1 of the algorithm produces the table shown in Figure 4.6.

Obj/Weight	0	1	2	3	4	5	6	7	8	9	10	11	12	13	14
1	0	0	7	7	7	7	7	7	7	7	7	7	7	7	7
2	0	8	8	15	15	15	15	15	15	15	15	15	15	15	15
3	0	8	8	15	15	15	22	22	29	29	29	29	29	29	29
5	0	8	8	15	15	20	22	22	29	29	34	34	34	34	34
5	0	8	8	15	15	20	22	25	29	30	34	34	39	39	44
6	0	8	8	15	23	23	30	30	35	37	40	44	45	49	49

Table 4.2. *The table obtained from phase 1 of the algorithm*

The final cell in the table (bottom right) reads 49, which determines the total maximum value.

By applying phase 2a of the algorithm, we determine that $j = 13$, i.e. the value of the maximum weight.

Following phase 2b, we obtain the indices of the columns:

i (no. col)	6	4	3	2	1

Table 4.3. *Phase 2b*

They provide the list of objects contained in the knapsack:

i (no. col)	6	4	3	2	1
object	F	D	C	B	A

Table 4.4. *Phase 2b, objects and the knapsack*

We can check that the total value of the objects is 49 and that their total weight is 13.

Value: 15 (F) + 5 (D) + 14 (C) + 8 (B) + 7 (A) = 49

Weight: 3 (F) + 2 (D) + 5 (C) + 1 (B) + 2 (A) = 13

4.3. Stochastic process

Before continuing, it is important to define the notion of stochastic processes that we will encounter later in this chapter.

Stochastic processes model systems whose behavior cannot, in part, be predicted. They have many applications, particularly for managing transport networks, managing traffic, electric systems, telecommunications, signal processing, filtering, finance, the economy, etc.

A stochastic process is a **family** $\{V_t, t \in E\}$ of random variables defined over a space of common probability. The index t is generally time. The process is said to be **continuous** if $E = [0, \infty[$ and **discrete** if $E = \{0, 1, 2, 3, ..., n\}$.

A stochastic process is said to be **Markovian** if, with respect to a present moment t, its future evolution does not depend on its past.

4.4. Markov chains

If, with respect to a physical or mathematical process, the following elements can be declared:

– it can have n possible states;

– at a moment t, it is in one and only one of its states;

– as time passes, the probability that it will be in a precise state only depends on its status at moment t-1.

Then this process is a **Markov chain**, named after its inventor A. Markov[4]. It can also be called a stochastic **Markovian** process.

A Markov chain is generally associated to a **transition matrix**, which describes the states of the systems at the different moments in the stages of the process.

If we consider the set of the states to be $E - \{e_1, e_2, ..., e_n\}$, the transition matrix M is a square matrix of dimensions $n \times n$ whose numbers are the probabilities p_e that define the states i of the process at moments t.

4.4.1. Property of Markov chains

A Markov chain can have one or a number of properties that give it specific functions, which are often used to manage a concrete case.

4.4.1.1. Absorbing chain

A chain can be **absorbing** when one of its states, called the **absorbing state**, is such it is impossible to leave once it has been entered.

The chain is absorbing if and only if one of its states is absorbing and if from a non-absorbing state we can reach an absorbing state.

The non-absorbing states in a chain are said to be **transient** or **transitory states**. It is worth noting that this state is not always returned to.

---

4 Andrei Markov, 1856–1922, Russian mathematician, member of the St Petersburg Academy of Sciences, worked on the theory of probabilities and stochastic calculation.

When a state is always returned to, at the end of a mean infinite time, it is said to be **null recurrent**.

When a state is returned to infinite times, at intervals of finite times, on average, it is said to be **positive recurrent**.

4.4.1.2. Irreducible chain

A Markov chain is said to be **irreducible** or **ergodic** when all its states communicate i.e. when for every couple of states (e_i, e_j), the probability of going from one to the other is strictly positive.

This property is easy to see on a transition graph (see section 4.4.2.3). Each couple of vertices is linked by one or a number of edges and once we have arrived on one of these vertices, we can leave (probability less than 1).

In the event the chosen vertex cannot be left, it is in an absorbing state and the chain is thus not irreducible.

It can thus be deduced that every absorbing Markov chain is not irreducible.

4.4.1.3. Homogeneous chain

A time-**Homogeneous** Markov chain is a Markov chain whose probability of transitioning is independent of time i.e. if E is a finite space and has countable states, n the moment the chain finds itself on a state, x and y of the states in E, we have:

$$\mathbb{P}(X_{n+1} = y | X_n = x) = \mathbb{P}(X_1 = y | X_0 = x)$$

4.4.1.4. Regular chain

A Markov chain is **regular** if we can go from any state towards any other state following a fixed number of steps, the starting state being independent.

Note that a regular chain is irreducible.

4.4.1.5. Reversible chain

A Markov chain of transition matrix M is **reversible** in relation to a probability π, if for all $x, y \in E$ we have:

$$\pi(x)M(x, y) = \pi(y)M(y, x)$$

4.4.2. *Classes and states of a chain*

Considering a transition matrix x, $y \in E$ of a Markov chain, a state j can be accessed from a state i if there is a path from i to j. The following property can therefore be defined:

the state j can be accessed from i means that $n \geq 0$ as long as $M^n_{ij} > 0$.

If the states i and j can be accessed from each other, they are said to be communicating and have the following property:

$n \geq 0$ and $m \geq 0$ as long as $M^n_{ij} > 0$ and $M^m_{ij} > 0$.

The relation enabling two states to communicate is an **equivalence relation** on which **classes** can be defined.

4.4.2.1. *Recurrent and transitory classes*

A class is said to be **recurrent** if it has a vertex without a successor in the transition graph.

A recurrent class comprising just one state is absorbing.

A class is said to be **transitory** if it is not recurrent.

4.4.2.2. *Period*

For a state i of a Markov chain, there is a period t which is equal to the GCD[5] of all the n for which $M^n_{ij} > 0$. The state i is **aperiodic** if $t = 1$ ($P_{ii} > 0$) and **periodic** if $t > 1$.

4.4.2.3. *Summary*

Let us consider the transition graph G of a Markov chain in Figure 4.6:

This graph can be said to have three classes: {A, B, C, F}, {D, E}, {G, H}.

The first is a transitory class and the two others are recurrent classes.

---

5 Greatest Common Divisor.

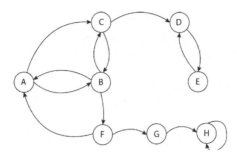

Figure 4.6. *The transition graph G*

Let us calculate the periods of states A and E:

For *A*, let us determine the length of each of the possible paths: *A, C, B, A* has a length of *3*; *A, B, F, A* has a length of *3*; *A, B, A* has a length of *2*; *A, C, B, F, A* has a length of *4*.

$$t(A) = PGCD(2, 3, 4) = 1$$

For *E*, there is only the path *D, E, D* which has a length of 2. The period is therefore:

$$t(E) = 2$$

A is an aperiodic state and *E* is a periodic state.

4.4.3. *Matrix and graph*

Let us consider the following example:

Jeanne, Pierre, François and Marie are playing frisbee[6] following a well-defined sequence of action. When Jeanne has the frisbee she throws it once every four times to Pierre, twice to Marie and once to François. If Pierre, the most technically able of the group, has the frisbee, he throws it once every five times back to himself, two times out of five to Marie, once to François

---

6 A frisbee is a plastic disc with turned-in edges, manufactured by Walter Morrison, a Yale student in 1948. He was inspired by William Russel Frisbie, a baker who wrapped up his cakes in steel molds of the same shape that students threw to each other at the end of meals. Nowadays, frisbee is a sport with a number of disciplines.

and once to Jeanne. Marie throws it twice every four times to Jeanne, once to François and once to Pierre. Finally François throws it three every six times to Pierre, twice to Marie and once to Jeanne.

The transition matrix is as follows:

$$M = \begin{pmatrix} 0 & 1/4 & 2/4 & 1/4 \\ 1/5 & 1/5 & 2/5 & 1/5 \\ 2/4 & 1/4 & 0 & 1/4 \\ 1/6 & 3/6 & 2/6 & 0 \end{pmatrix}$$

The transition matrix can be represented by a digraph or a transition graph whose summits are the states of the process and the edges are the ordered couples of the states.

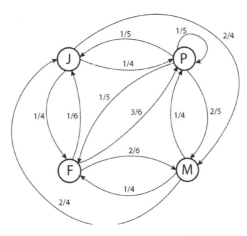

Figure 4.7. *The digraph corresponding to our example*

If we take a close look at the matrix in our example, it can be seen that the total elements in each line is equal to 1 and that each element therefore varies between 0 and 1. A matrix of this type said to be a **stochastic matrix**.

4.4.4. *Applying Markov chains*

A car hire company has three agencies: one in Paris with 110 cars, one in Lyon with 100 cars and one in Marseille with 80 cars. A client can hire a car in any town and take it back to this town or to any other town. After being in business for several months, this company can formalize its monthly figures, which are collected in the table below.

	Arriving at PARIS	Arriving at LYON	Arriving at MARSEILLE	Missing
Leaving from PARIS	40%	30%	20%	10%
Leaving from LYON	30%	45%	20%	7%
Leaving from MARSEILLE	15%	25%	45%	15%

Table 4.5. *Monthly figures for the hire company*

During their journey, some cars are stolen or involved in an accident. These cars are considered to be missing.

To make future investments, the management decided to analyze its car park. After three months how many cars are still available to clients?

To answer this question, we start by determining the states that are likely to take cars.

They can be in Paris, Lyon, Marseille or can be missing, i.e. there are 4 states thus giving us the possibility to create the following transition matrix for the first month.

$$M = \begin{pmatrix} 0.4 & 0.3 & 0.2 & 0.1 \\ 0.3 & 0.45 & 0.2 & 0.05 \\ 0.15 & 0.25 & 0.45 & 0.15 \\ 0 & 0 & 0 & 1 \end{pmatrix}$$

For the second month, we get:

$$(110 \quad 100 \quad 80 \quad 0) \begin{pmatrix} 0.4 & 0.3 & 0.2 & 0.1 \\ 0.3 & 0.45 & 0.2 & 0.05 \\ 0.15 & 0.25 & 0.45 & 0.15 \\ 0 & 0 & 0 & 1 \end{pmatrix} = (82 \quad 95 \quad 76 \quad 27)$$

That means that the company has 82 cars in Paris, 95 cars in Lyon, 76 cars in Marseille and that 27 are missing, totaling 253 cars.

For the third month, we get:

$$(110 \quad 100 \quad 80 \quad 0)\begin{pmatrix} 0.4 & 0.3 & 0.2 & 0.1 \\ 0.3 & 0.45 & 0.2 & 0.05 \\ 0.15 & 0.25 & 0.45 & 0.15 \\ 0 & 0 & 0 & 1 \end{pmatrix}^2 =$$

$$(110 \quad 100 \quad 80 \quad 0)\begin{pmatrix} 0.28 & 0.305 & 0.23 & 0.185 \\ 0.285 & 0.3425 & 0.24 & 0.1325 \\ 0.2025 & 0.27 & 0.2825 & 0.245 \\ 0 & 0 & 0 & 1 \end{pmatrix} =$$

$$(75.5 \quad 89.4 \quad 71.9 \quad 53.3)$$

At the end of the 3rd month the company has 235 cars $(75 + 89 + 71)$.

4.5. Exercises

4.5.1. *Exercise 1: Levenshtein[7] distance*

This distance, often denoted as LD, measures the similarity between two chains of characters, whether they be numeric, alphabetic or alphanumeric. We start from a source chain S to arrive at a target chain C. LD is equal to the minimum number of transformations that must take place to transform S into C. There are three types of transformations:

– replacing a character of S with a character of C;

– removing a character of S;

– adding a character of C in S.

COMMENT.– The value of each transformation is equal to 1, unless a character is replaced by an identical one.

Create an algorithm using dynamic programming to calculate the LD between S and C. You will have to use a matrix (double entry table).

---

7 Vladimr Levenshtein, 1935, a Russian scientist specializing in code theory, proposed the calculation for this mathematical distance in 1965. It is also known as edit distance or dynamic time deformation.

4.5.2. Exercise 2

A mouse is moving around the maze depicted in Figure 4.9 which contains five cells A, B, C, D and E. At fixed intervals of time, the mouse moves from one cell to another. When the mouse is in a cell that has a number of access points, it chooses one of these access points with an equal probability. The mouse's decision of where to move next is never influenced by its previous action.

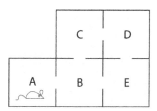

Figure 4.8. *The maze for exercise 2*

Questions:

1) Represent the graph and the transition matrix M

2) At each moment t, the mouse is in cell N, let us consider two paths C_{AE} and C_{AD} formed of a set of couples (t, N):

$$C_{AE} = \{(t_0, A), (t_1, B), (t_2, C), (t_3, D), (t_4, E)\}$$

$$C_{AB} = \{(t_0, A), (t_1, B), (t_3, E), (t_4, B), (t_5, C), (t_6, D), (t_7, B)\}$$

Calculate the probability for each of these paths.

3) Calculate the probability of the mouse, which is initially in cell A, of reaching cell E in 2 or 3 steps and reaching cell D in 3 steps.

4.5.3. Exercise 3: Ehrenfest[8] model

We draw balls out of bags. There are four balls in two bags. Each time we draw a ball, we pick one ball out of four at random, with equal probability, and then put it in the other bag.

---

8 Paul Ehrenfest, 1880–1933, Austrian physicist, who with his wife Tatiana was a major player in launching statistical mechanics.

Questions:

1) If T_n is the number of balls in the first bag after n draws, represent the transition graph of the Markov chain associated to this process.

2) Determine the transition matrix.

3) Is the chain irreducible or ergodic?

4) Is the chain regular?

4.5.4. *Solutions for exercise 1*

For this algorithm, we use the following elements:

– u: length of the chain S ;

– v: length of the chain C ;

– i and j: iterators ;

– $LD[i, j]$: matrix containing the indices attached to S and C with dimensions $u + 1$ and $v + 1$;

– $S[i]$: th character of the chain S ;

– $C[i]$: th character of the chain C ;

– $cost$: value (0 or 1) of the similarity between a character of S and a character of C.

```
for i = 0 to u do
        LD[i, 0] = i
end for
for j = 1 to v do
        LD[0, j] = j
end for
for i = 1 to u do
        for j = 1 to v do
                if S[i] = C[j] then
                        cost=0
                otherwise
                        cost=1
                end if
                LD[i, j] = LD[i - 1, j] + 1
                if LD[i, j - 1] + 1 < LD[i, j] then
                        LD[i, j] = LD[i, j - 1] + 1
```

```
            end if
            if LD[i - 1, j - 1] + cost < LD[i, j] then
                    LD[i, j] = LD[i - 1, j - 1] + cost
            end if
        end if
end if
return LD [u, v]
```

The first test detects a similarity between a character of S and a character of C.

Between the first and the second test, the code removes a character.

The second test inserts a character.

The third test replaces a character.

At the end of the algorithm, $LD[u, v]$ contains the LD value (number of transformations).

4.5.5. *Solutions for exercise 2*

Question 1:

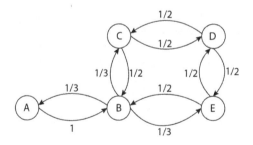

Figure 4.9. *The graph for exercise 2, question 1*

$$M = \begin{pmatrix} 0 & 1 & 0 & 0 & 0 \\ 1/3 & 0 & 1/3 & 0 & 1/3 \\ 0 & 1/2 & 0 & 1/2 & 0 \\ 0 & 0 & 1/2 & 0 & 1/2 \\ 0 & 1/2 & 0 & 1/2 & 0 \end{pmatrix}$$

Question 2:

$$C_{AE} = \{(t_0,\ A),\ (t_1,\ B),\ (t_2,\ C),\ (t_3,\ D),\ (t_4,\ E)\}$$

$$P_{C_{AE}} = 1 \times \frac{1}{3} \times \frac{1}{2} \times \frac{1}{2} = \frac{1}{12}$$

$$C_{AB} = \{(t_0,\ A),\ (t_1,\ B),\ (t_3,\ E),\ (t_4,\ B),\ (t_5,\ C),\ (t_6,\ D),\ (t_7,\ B)\}$$

$$P_{C_{AB}} = 1 \times \frac{1}{3} \times \frac{1}{2} \times \frac{1}{3} \times \frac{1}{2} \times 0 = 0$$

Question 3:

To get from A to E in 2 steps there is one path $C = \{(t_0,\ A),\ (t_1,\ B),\ (t_4,\ E)\}$ therefore

$$P_C = 1 \times \frac{1}{3} = \frac{1}{3}$$

It is not possible to get from A to E in 3 steps therefore $P = 0$

To get from A to E in 3 steps there are 2 possible paths:

$C_1 = \{(t_0,\ A),\ (t_1,\ B),\ (t_2,\ C),\ (t_3,\ D)\}$ et $C_2 = \{(t_0,\ A),\ (t_1,\ B),\ (t_2,\ E),\ (t_3,\ D)\}$ therefore:

$$P = P_{C_1} + P_{C_2} = 1 \times \frac{1}{3} \times \frac{1}{2} + 1 \times \frac{1}{3} \times \frac{1}{2} = \frac{1}{3}$$

4.5.6. *Solutions for exercise 3*

Question 1:

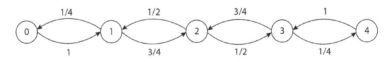

Figure 4.10. *The graph of the Erhenfest model*

Question 2:

$$
M = \begin{pmatrix}
0 & 1 & 0 & 0 & 0 \\
1/4 & 0 & 3/4 & 0 & 0 \\
0 & 1/2 & 0 & 1/2 & 0 \\
0 & 0 & 3/4 & 0 & 1/4 \\
0 & 0 & 0 & 1 & 0
\end{pmatrix}
$$

Question 3:

From any of the states there is a path towards another. The chain is therefore irreducible or ergotic.

Question 4:

The chain is not regular. If we start from state 0 to go toward states 2 or 4, we will find and an even number of draws, whereas to go toward 1 or 3 the number of draws will be uneven. The element $(M^t)_{ij}$ is equal 0 when $i + j + n$ is uneven.

Scheduling with PERT and MPM

5.1. The main concepts

The aim of scheduling methods is to plan a set of tasks or activities so as to meet one or a number of final objectives described in the scope statement attached to a project.

Tasks can be multiple and varied. They implement different resources, workforces, machines, materials and raw materials, associated with a global calendar in which the period for carrying out the work within the enterprise's organization is defined with a start date and an end date. The conditions of the deadlines, resources and costs must be optimized to determine maximum profitability.

Implementing a system, a new organization, a workshop, an assembly line, a manufacturing chain, the construction of a building, the maintenance of a system, the creation of a prototype, the implementation of a promotional or advertising campaign are all examples of projects that may use scheduling methods.

When a product or a job has to be carried out in a given amount of time, we often find ourselves faced with often complex scheduling problems. Sequential or parallel different activities, different constraints, whether they are linked to resources, time, finance or anything else, as well as the risks that have to be taken into consideration rapidly form an imbroglio that is not always easy to untangle. Tasks are interdependent and it becomes difficult to see the overall picture, as any modification leads to disruptions and reactions within the whole process being realized. Moreover, large-scale projects must be divided into subprojects, each of which involving one or a number of processes, in turn requiring specific scheduling, while maintaining a close relationship with its neighbors.

The scheduling of a project can be represented with a graph, which can take a number of forms, depending on the method used. There are two very widely used methods: **Program Evaluation and Review Technique or Program Evaluation Research Task – Critical Path Method (PERT-CPM)** and *Méthode des potentiels Métra* **(MPM)**.

PERT or MPM graphs can easily be converted into a Gantt chart.

5.2. Critical path method

PERT and MPM methods are based on dividing the project into tasks or elementary activities of estimated durations, which are logically interconnected. This analysis phase will enable a graph, also known as a network, to be constructed. Determining the planning, also known as scheduling, will always seek to find a critical path i.e. a path that will determine the total duration of the project. Then resources and constraints required for realising the project will be implemented such as to calculate costs. This stage may question the original planning and require the graph to be adapted to improve or resolve the invisible risks when there are no resources.

In the remainder of this chapter, the author will use the terms tasks and activities interchangeably as these two terms are perfect synonyms within the domain of scheduling methods.

In the established graph, a set of interconnected tasks or activities form the critical path. In each graph, there is always at least one critical path. The tasks on this path are called **critical tasks**.

It should be noted that a scheduling or planning graph has precise characteristics. It is connected, without cycle and valued.

Critical tasks benefit from a great importance within the graph because they have a direct impact on the overall duration of the project and therefore on its end or start date (**retroactive planning**). Consider that within a graph of medium or large importance (above 50 tasks), 10–15% of activities are critical.

5.3. Precedence diagram

Regardless of the method chosen, to make it easier to construct the graph, we define a **precedence table**.

The table can have different aspects such as a double-entry matrix.

	A	B	C	D	E	F	G
A		X	X				
B					X		
C				X		X	
D					X		
E							X
F							X
G							

Figure 5.1. *A matrix*

The matrix is read from line to column.

This way of doing things is often called Cartesian due to the shape of the table.

For example:

– task A is the predecessor of tasks B and C;

– it can also be said that tasks B and C have task A as an immediate predecessor;

– task B has task E as a successor or next task.

We can also find a table setting out a list of tasks to which durations and **immediate predecessors** are associated.

Task	Duration	Predecessor
A	5	–
B	4	A
C	6	A
D	2	C
E	7	B; D
F	5	C
G	2	E; F

Table 5.1. *Predecessors*

Tasks which do not have predecessors are the tasks starting the project. They are called **start tasks**.

Finally, a more complex table including tasks, durations, immediate predecessors and successors is another possible presentation.

Task	Duration	Predecessor	Successor
A	5	–	B; C
B	4	A	E
C	6	A	D; F
D	2	C	E
E	7	B; D	G
F	5	C	G
G	2	E; F	–

Table 5.2. *Predecessors-successors*

Tasks that do not have any successors are tasks ending the project. They are called **end tasks**.

COMMENT 5.1.– In all precedence tables, the immediate qualifier indicates that only the predecessors or successors linked to the ongoing task are mentioned.

Tasks are often symbolized by the letters: A, B, ...,Y, Z, AA, AB, ..., ZZ,... so they can be easily handled and will not overload the table, the matrix and later the graph. Next to their symbol, a description or a more explicit literal designation is often added.

To sum up, the predecessor table brings together the three standard constraints linked to task scheduling:

– **Potential constraints** are based on the successive aspect of tasks (a task must not be performed before another) or on the time the task is performed (a task must not start before a given date, a task must finish before a given date).

– **Cumulative constraints** take into account the limits imposed by production (machines, workforce, raw materials, costs, etc.).

– **Disjunctive constraints** are based on the non-parallelism or non-simultaneity of two or more tasks being performed.

Once we have the precedence table, we can create the graph (network) via the precedence diagramming method (**PDM**) which will create a PERT-CPM or MPM precedence diagram.

5.4. Planning a project with PERT/CPM

5.4.1. *History*

The Program Evaluation and Review Technique or Program Evaluation Research Task (PERT) method was developed between 1956 and 1958 upon request from the special projects' office at the American marines which commissioned the research to strategy consultancy and service firm Booz Allen and Hamilton[1].

The main objective was to establish the construction program for the Polaris ballistic nuclear missile[2] while keeping the method developed coherent with a method already in place in the marines, "**Milestone reporting**".

---

1 American firm specializing in management consultancy whose headquarters are based in Tysons Corner Virginia.
2 Strategic missile constructed by Lockheed for the US navy and Royal Navy. It was operated for the first time in July 1960.

The necessary calculations were made on the IBM computer at the calculation center (NORC[3]) of the marines located at Dalhgren in Virginia.

The military was not the only organization to be interested in these types of methods. Private industry, particularly the Du Pont de Nemours firm also worked on scheduling, and in 1957 Walker and Kelley[4] developed CPM which would be used for the first time in 1958 to construct a chemical factory and then in 1959 in the context of Du Pont de Nemours factories in Louisville.

The development and calculations were conducted on a UNIVAC-1 type computer.

The main differences between PERT and CPM lie particularly in the graphic representation, the management of task duration and the association of duration and cost, which appeared in CPM.

After 1958, there was a proliferation of planning methods, such as Program Evaluation Procedure (PEP) from the US Air Force. Countless programs like scheduling and control by automated network systems (SCANS) and least cost estimating and scheduling (LESS) were developed for the computers available at the time.

As time went by, the different trends constantly moving around PERT and CPM started to blur until there was just one method known as PERT or PERT-CPM.

This method would not be known in Europe until the 1960s.

5.4.2. Methodology

5.4.2.1. Formulating the graph

The PERT method uses three types of symbols:

– vertices (or nodes) designating the steps;

---

3 IBM Naval Ordnance Research Calculator. One of the first supercalculators, built in 1954 and used until 1968.
4 Morgan R. Walker from the American firm Dupont, James E. Kelley from the American firm Remington Rand. They developed CPM in the late 1950s.

– edges (arrows) in a solid line representing the task or activity;

– edges (arrows) in dotted lines representing the fictive task or activity (sequential link between two steps, obligatory for certain scenarios such that the graph remains coherent, see below).

A task is a vector defined by vertices indicating its start and end. The length of the vector varies and does not depend on the duration of the task. The edge symbolizing the task is associated with one or more letters codifying the task. This code has an often indexed value representing the duration of the task according to a unit (second, minute, hour, day, week, and so on) provided in a key in the graph.

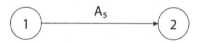

Figure 5.2. *A codified task or activity A, with a duration equal to 5, defined by vertices 1 and 2*

The vertices are numbered successively from 1 at the start to the finish. The number of the vertex (step) marking a task is finished must always be higher than the number of its start vertex.

In the PERT graph, time passes from left to right. A finish step is always to the right of a start step.

In a PERT graph, there are two types of particular tasks:

– the "start" task, starting the graph on the far left and which must, by definition, precede all other tasks;

– the "finish" task, finishing the graph on the far right and which must be performed after all other tasks.

Tasks can be assembled in several ways, thus creating dependency rules:

– **sequential or successive task**: a task cannot start before another has finished. There is a successor task and a predecessor task;

– **simultaneous or parallel tasks**: two or more tasks start at the same time. Their run-times overlap;

– **convergent tasks**: two or more tasks meet at one step;

– **divergent tasks**: two or more tasks start at the same time following the same step.

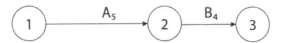

A and B are sequential or successive

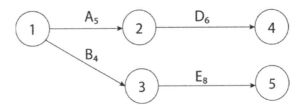

D and E are simultaneous or parallel

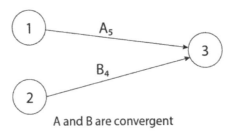

A and B are convergent

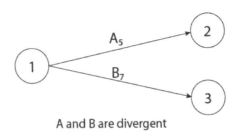

A and B are divergent

Figure 5.3. *The different ways tasks can be assembled in a PERT graph. Above A is the predecessor to B and B is the successor to A*

Playing with the different assembly combinations creates complex dependencies.

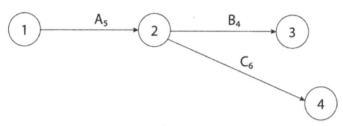

Tasks B and C can only start if task task A is finished

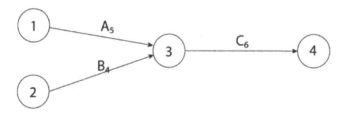

Task C can only start if tasks A and B are finished

Figure 5.4. *Two examples of the complex dependencies of a number of tasks*

When tasks overlap, there are a number of possible approaches for how to overlap them.

– start-start: a task can start once another task has started;

– finish-finish: a task can finish once another task has finished;

– start-finish: a task can finish once another task has started;

– finish-start: a task can start before another task finishes.

5.4.2.2. *Fictive task*

In a PERT graph, under certain conditions, we may need a special type of task: a fictive task which is drawn with a dotted line. Fictive tasks are there to express that a step cannot take place before another, even if there is no task between the two. Theefore, it avoids having two parallel edges (tasks) linking the same vertices (steps). Its duration is always null.

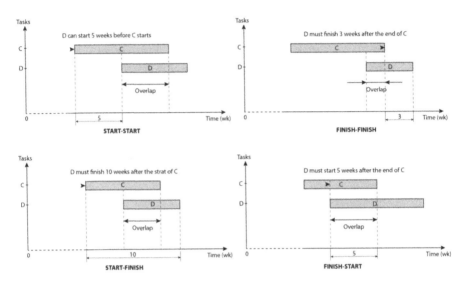

Figure 5.5. *Different approaches to overlap*

To better understand the principle of the fictive task, let us consider the relevant part of the graph in Figure 5.6.

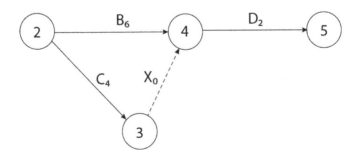

Figure 5.6. *A fictive task X within a graph*

Task D cannot start unless tasks B and C have finished. To express and meet this constraint without putting two parallel edges between step 2 (vertex 2) and step 4 (vertex 4), which is prohibited in a PERT graph, we add a fictive task (as a dotted edge) here called X, of a duration equal to 0.

It is possible to draw a graph by hand when there is a reasonable number of tasks. When there are a large number of tasks, i.e. above one hundred, we use software tools.

5.4.2.3. *Milestone tasks*

When the project involves a large number of tasks, it can be divided into clearly identified phases demarcated by **milestone tasks**. Milestones are often associated with operations such as checking, producing a document and holding a meeting.

A milestone always has a duration of null as it does not represent a job considered part of the project planning.

5.4.2.4. *Building rules*

To create a graph by hand, certain rules must be respected while it is being built so it remains clear and comprehensible:

– draw from left to right;

– do not cross edges (tasks) where possible;

– draw edges as a straight line where possible;

– avoid having edges that are too long;

– draw edges whose angles are as big as possible in relation to the edges;

– set a maximum number of fictive tasks. Do not have useless fictive tasks.

5.4.2.5. *Numbering the vertices*

The vertices in a PERT graph are all numbered sequentially from the start task to the finish task. To number the vertices, there is just one rule: the number of the step indicating the task is finished must always be higher than the number of its start task. There are often several different ways to number the vertices which respect to this constraint on the same graph.

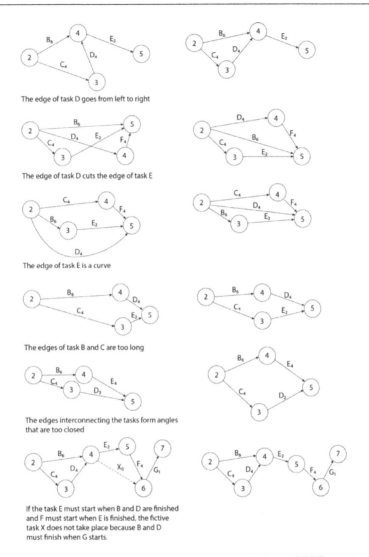

Figure 5.7. *The rules to be respected while building a PERT graph*

American mathematician Delbert Ray Fulkerson[5] developed the following simple method to number vertices:

– find the initial vertex (step) and number it 1;

_____

5 American mathematician, 14th August 1924–10th January 1976.

– remove all edges (tasks) whose origin is already numbered to create a new initial vertex;

– number the new initial steps(s) 2, 3, 4,… in any order;

– remove the edges whose origins are these vertices;

– continue this process until the finish vertex. The finish vertex does not precede any edge;

– when numbering is finished, a task can be identified by two values representing its origin and its extremity.

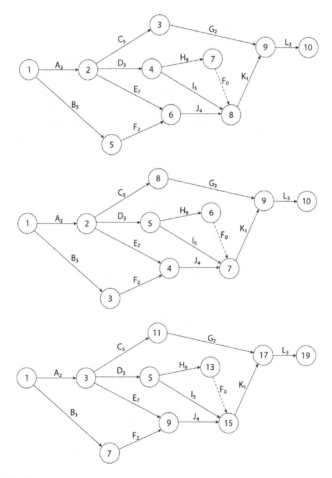

Figure 5.8. *Several examples of different ways of numbering graphs. In the top graph, task E, with a duration of 7, can be identified with the pair (2, 6)*

Numbering always goes up but its rate of increase may be more than 1 so that a new task can be inserted without the whole graph needing to be renumbered, for example from 2 on 2, from 10 on 10 etc.

5.4.2.6. *Earliest dates and latest dates*

To determine the critical path, once the graph has been created and numbered, the earliest dates and latest dates must be defined.

There are two types of start date:

– **earliest start**, which is the earliest date a task can start;

– **earliest finish**, which is the latest date a task can finish.

They are obtained by traversing the graph from left to right and by adding up (at each vertex) the duration of each of the tasks from the start date of the project. In the case of convergence, the highest value is chosen.

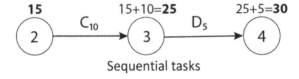

Sequential tasks

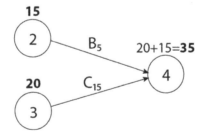

Convergent tasks (20+15=**35**>*15+5=20*)

Figure 5.9. *The different types of calculations for an earliest date in PERT graph*

As before, there are two types of latest dates:

– **latest start**, which is the latest date a task can start without the finish date of the project being moved;

– **latest finish**, which is the latest date a task can finish without the finish date of the project being moved.

They are obtained by traversing the graph from right to left (from the project finish step until its start step) by subtracting (at each vertex) the duration of each of the tasks. The project finish vertex has a latest date equal to its earliest date. In the case of convergence (divergence from left to right), the smallest value of the difference is chosen.

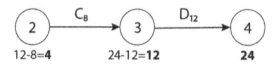

Sequential tasks

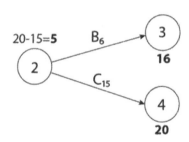

Convergent tasks from right to left,
divergent from left to right (**20-15=5**<16-6=10)

Figure 5.10. *The different types of calculations*
for a latest date in a PERT graph

5.4.2.7. Determining the critical path

The critical path passes through the vertices whose earliest and latest dates are equal. There could be several possible routes when graphs become complex.

If a graph has fictive tasks, a path that does not pass through any of these will be chosen if possible.

A modification to the duration of a critical task will have immediate consequences on the overall duration and therefore the finish date of a project. However, a modification to the duration of one or a number of non-critical tasks will not necessarily have an influence and the work may finish within the predicted timeframe if available slacks are respected (see below, section 5.6).

Exceeding the slacks will lead to the critical path being modified and therefore the tasks they comprise.

5.5. Example of determining at critical path with PERT

Let us consider the following (real) scenario:

We want to determine the scheduling for the installation of an open-air sound system.

It is composed of a mixing console, an amplification system, two loudspeakers systems (low, medium and high) facing the public to the front and two large speakers at the back. The operations to be performed follow a well-defined process and some can be carried out at the same time. The material is unloaded, the different elements are installed, the wiring is arranged and then there is a final test of the whole system.

The operations are as follows:

– transporting the material (120 min);

– unloading front and back speakers (20 min);

– unloading and installing amplification system and cables (15 min);

– unloading mixing console (20 min);

– wiring front speakers (10 min);

– installing cabling console and amplification system (15 min);

– assembling, putting together and connecting front speakers (25 min);

– wiring back speakers (20 min);

– test installation (15 min);

Some operations have been purposefully removed to simplify the example.

5.5.1. *From the scenario, creating the predecessor table*

Task	Description	Duration (min)	Predecessor
A	Transporting the material	120	–
B	Unloading front and back loudspeakers systems	20	A
C	Unloading and installing amplification system and cables	15	A
D	Unloading mixing console	20	A
E	Wiring front speakers	10	B
F	Installing cabling console and amplification system	15	C; D
G	Assembling, putting together and connecting front speakers	25	E; F
H	Wiring back speakers	20	B
I	Test installation	15	G; H

Table 5.3. *Tasks and predecessors*
for a sound system

5.5.2. *Creating the graph*

Draw the start task between 2 vertices (steps), then the successors by respecting the predecessors until the final vertex. On each edge we place the task code alongside its duration.

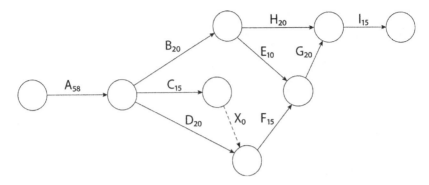

Figure 5.11. *The graph for our example relating to the precedence table. Note the presence of the fictive task X with a duration of 0*

In our example, A is the start task (it does not have a predecessor) and I is the finish task (it does not have a successor).

While creating this graph, task F encounters a problem as it has two predecessors C and D. As two specific vertices cannot be linked by 2 parallel edges, we decide to define a fictive task (dotted line). By definition, its duration is 0.

It is worth pointing out that it would have been possible to draw a fictive task from task D instead of fictive task from task C. This would have made no difference to our future calculations.

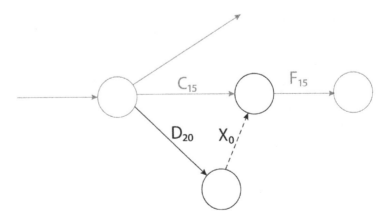

Figure 5.12. *Another possibility of where to draw fictive task X*

5.5.3. Numbering the vertices

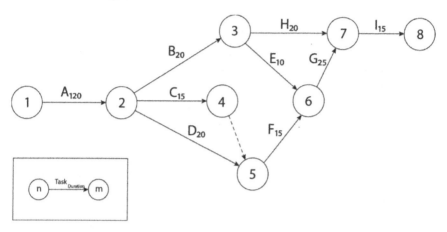

Figure 5.13. *The graph with its vertices (steps)*

5.5.4. Determining the earliest dates of each of the tasks

Here is the approach for calculating the earliest dates for our example (to be as clear as possible, a number is considered to be a vertex or step and a letter is considered to be an edge symbolizing a task or activity):

– we go to 1 and attribute it the earliest date value *0*;

– to go from 1 to 2 we pass through A with a duration of *120*. We add *0* + *120*, that is *120*, which becomes the earliest date of 2;

– from 2 we can reach 3, 4 and 5. We add the earliest date of 2 with the duration of B, that is *120 + 20*, giving us *140* for 3;

– to calculate the earliest date of 4, we add the earliest date of 2 with the duration of C, that is *120 + 15*, giving us *135* for 4;

– to calculate the earliest date of 5 we have two possibilities as 2 edges, formed by D and X, converge at 5. We can therefore add *120 + 20* and *135 + 0* deriving from 2 and 4 respectively. During a situation of convergence, we always choose the highest total. We get an earliest date of *140* on 5;

– to calculate the earliest date of 6 we have two possibilities as 2 edges, formed by E and F, converge at 6. We can therefore add *140 + 10* and *140 +*

15 deriving from 3 and 5 respectively. Once again there is a convergence and we always choose the highest total. We get an earliest date of *155* on 6;

– to calculate the earliest date of 7, there are once again two possibilities as G and H converge. We can add *155 + 25* or *140 + 20* deriving from 6 and 3 respectively. We choose the highest total, *155 + 25* that is *180* as the earliest date for 7;

– finally, 8 must be reached via I to which we attribute the calculation *180 + 15* that is *195*.

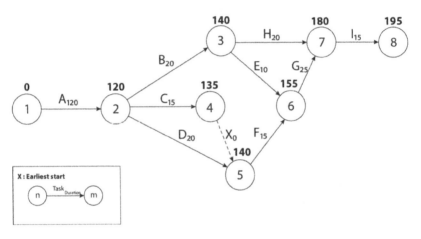

Figure 5.14. *PERT graph with earliest dates calculated and attributed above each vertex*

5.5.5. Determining the latest dates of each of the tasks

Here is the approach for calculating the latest dates for our example:

– we go to 8 and attribute the value of its start date as its latest date, that is *195*;

– to go from 8 to 7, we pass through I with a duration of *15*, we calculate the difference *195 – 15*, that is *180*, which we attribute to 7;

– to go from 7 to 6, we pass through G with a duration *25*, we calculate the difference *180 – 25*, that is *155*, which becomes the latest date of 6;

– to get to 3, we can pass through E or H, indicating a convergence. There are two possible calculations: *155 – 10* by passing through E or *180 – 20* by passing through H. We choose the smallest difference, *145*;

– to get to 5 from 6, we pass through F with a duration of *15*, we calculate the difference *155 – 15*, that is *140*, which we attribute to 5;

– to get to 4 from 5, we pass through X with a duration of *0*, *140* is attributed to 4;

– to get to 2, once again there is a convergence. There are three possible paths: from 3 passing through B; from 4 passing through C; or from 5 passing through D. The calculations are *145 – 20*, *140 – 15* or *140 – 20*, respectively. We choose the smallest difference, that is *120* as the latest date of 2;

– we can then reach 1 via A for which the latest date will be the difference *120 – 120*, that is *0*.

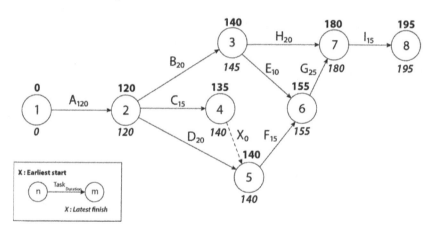

Figure 5.15. *PERT craft with latest dates calculated and displayed below each vertex*

Note that the earliest dates and latest dates of the 1st vertex in the graph are equal (in our example *0*). If during calculations, we get different values, at least one of the calculations forwards or backwards is incorrect.

Here is the table collecting together the earliest start dates and latest finish dates for each of the vertices in our example:

Start vertex	Earliest start	Task	Duration	End vertex	Latest finish
1	0	A	120	2	120
2	120	B	20	3	145
2	120	C	15	4	140
2	120	D	20	5	140
3	140	E	10	6	155
3	140	H	20	7	180
4	135	X	0	5	140
5	140	F	15	6	155
6	155	G	25	7	180
7	180	I	15	8	195

Table 5.4. *Earliest starts and latest finishes*

COMMENTS 5.2.– *Start and end vertices always have the same earliest and latest dates. When calculating a latest date, there can be no negative value.*

The earliest date of the start vertex is not always equal to 0. The graph in question may be a subset (subproject) of another more ambitious project. In this case, the earliest date will have a value higher than 0 (which is also equal to its latest date, respecting the first comment made above).

5.5.6. Determining one or more critical paths

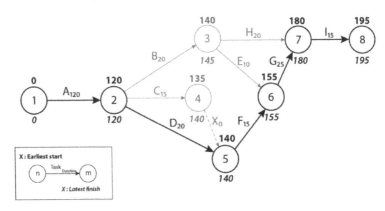

Figure 5.16. *The critical path in our example*

In our example, the total maximum duration of the project will be 195 min. The critical path is the route 1, 2, 5, 6, 7, 8 and the critical tasks are A, D, F, G and I.

A modification to the duration of one of these critical tasks will change the overall duration of the project and therefore its finish date. A change to the duration of one of the non-critical tasks B, C, E and H (we do not consider the fictive task X) will not necessarily have an impact if the available slacks are respected.

It can be noted that a critical task (edge) is defined between two steps (vertices), each of which have an earliest date and a latest date. We can therefore say that a task is defined by four values which can redefine it. For the step on the left (start of the task), we have the earliest start date and the latest start date and for the step on the right (end of the task), we have the earliest finish date and the latest finish date.

Figure 5.17. *Earliest start, latest start, earliest finish and latest finish available for a task on the PERT graph*

For each of the tasks, whether critical or not, we get:

Earliest finish = Earliest start + Task duration

Latest start = Latest finish – Task duration

5.6. Slacks

There are several types of slacks which give different information:

– **total slack**;

– **free slack**;

– **independent slack**.

Slacks are often considered hesitation intervals. We often say that one of the characteristics of the critical path is that it has no hesitation intervals.

5.6.1. *Total slack*

The total slack attached to a task is a delay for a task which does not delay the project as a whole provided the task in question starts at its earliest date. It provides the degree of criticality of the path.

Total slack = Latest finish – Earliest start – Task duration

5.6.2. *Free slack*

The free slack *(or free float)* attached to a task is the delay for a task that does not delay the earliest state date of one of its successors, provided that the task in question starts at its earliest date.

If the free slack is exceeded, some of the successive tasks will be delayed. The impact on the duration of the project varies (though it is often null).

Free slack = Earliest start of the successive task – (Earliest start + Task duration)

or

Free slack = Earliest start of the successive task – Earliest finish

The free slack of the last task is always 0.

5.6.3. *Independent slack*

The independent slack *(or independent float)* attached to a task is a delay to the task that does not modify the duration of the project regardless of start date of the task in question.

Independent slack = Earliest finish - Latest start – Task duration

A negative certain slack is considered equal to 0.

5.6.4. *Properties*

Critical tasks always have null slack.

The free slack is always less than or equal to the total slack.

A non-critical task linking two critical tasks has a total slack equal to its free slack.

When several successive non-critical tasks link two critical tasks, there can be a difference between the total slack and the free slack.

5.7. Example of calculating slacks

The table collecting the calculations of slacks from the previous example (see section 5.5.1) can be seen below.

In this table, the lines in bold represent critical tasks.

Task	Start step	Finish step	Duration	Earliest start	Latest start	Earliest start	Latest finish	Total slack	Free slack	Independent slack
A	**1**	**2**	**120**	**0**	**0**	**120**	**120**	**120-0-120=0**	**120-0-120=0**	**120-0-120=0**
B	2	3	20	120	125	140	145	145-120-20=5	140-120-20=0	140-125-20=5 => 0
C	2	4	15	120	125	140*	140*	140-120-15=5	140-120-15=5	140-125-15=0
D	**2**	**5**	**20**	**120**	**120**	**140**	**140**	**140-120-20=0**	**140-120-20=0**	**140-120-20=0**
E	3	6	10	140	145	150	155	155-140-10=5	155-140-10=5	150-145-10=0
F	**5**	**6**	**15**	**140**	**140**	**155**	**155**	**155-140-15=0**	**155-140-15=0**	**155-140-15=0**
G	**6**	**7**	**25**	**155**	**155**	**180**	**180**	**180-155-25=0**	**180-155-25=0**	**180-155-25=0**
H	3	7	20	140	160	160	180	180-140-20=20	180-140-20=20	160-160-20=20 => 0
I	**7**	**8**	**15**	**180**	**180**	**195**	**195**	**195-180-15=0**	**0**	**195-180-15=0**
X	*4*	*5*	*0*	*135*	*140*	*135*	*140*	–	–	–

*The fictive task X must be taken into consideration and step 5 must be considered as being the finish step of task C.

Table 5.5. *Calculation of slacks*

5.8. Determining the critical path with a double-entry table

To determine the critical path from a PERT graph, we can use a method based on a table also known as a matrix.

5.8.1. *Creating a table from our example*

We define a table comprised of N+2 columns and lines, i.e. if we take up the aforementioned example (see section 5.5.3), 10 lines × 10 columns.

⇒	1	2	3	4	5	6	7	8	*ES*
1									
2									
3									
4									
5									
6									
7									
8									
LF									

Table 5.6. *Double-entry matrix, empty, for our example*

The diagonal and the part below it are defined.

There should be no value below or on this diagonal.

If this happens, there is an error in the graph report in the table or the numbering of the vertices in the graph does not respect the conditions described in section 5.4.2.5.

The first line and the first column contain the numbers of the vertices. The last column contains the earliest start (*ES*) and the final line contains the latest finish (*LF*) of each task. The graph is read from line to column.

5.8.2. *Filling in the table*

We place the duration of each task in the respective cells. We start with vertices 1 and 2, which define task A for a duration of *120* which we put in the matrix, and then vertices 2 and 3, which define task B with a duration of *20*. We continue in this way until the final task I and obtain the table below. If there are fictive tasks in the graph, they must always be included.

$\Rightarrow$	1	2	3	4	5	6	7	8	*ES*
1		120							
2			20	15	20				
3						10	20		
4					0				
5						15			
6							25		
7								15	
8									
LF									

Table 5.7. *Double-entry matrix, filled in with the duration of tasks*

As for the previously seen method, we start by determining the earliest dates, then the latest dates until we define the project duration and the critical path.

5.8.3. *Earliest start dates*

The algorithm below is used to determine these dates:

– we place *0* (or any other value if the graph being drawn is a subproject) on line 1 in column *ES*;

– we look for the value on the same line, against the diagonal, to which we add the value of line 1 column *ES* and place it in the line below;

– remaining on this line, we will look for the value against the diagonal. Three cases are possible:

- the value is alone in the column,

- several values exist in the column,

- the value does not exist, but other values are present in the column,

– in the first case, we add it to the previously found value;

– in the second and third cases, we add each value to the previously found values line by line and choose the highest value. Careful: an inexistent value is never considered equal to 0. When there is no value, no calculation is possible;

– we continue in this way until we reach the final line.

Using this algorithm on our example gives the following calculations:

– Line 1 : *ES = 0*

– Line 2 : *ES = 0 +120 = 120*

– Line 3 : *ES = 120 + 20 = 140*

– Line 4 : Case c => *ES = 120 + 15 = 135*

– Line 5 : Case b => We have *135 + 0 = 135* or *120 + 20, ES = 140* (the highest total)

– Line 6 : Case b => We have *140 + 15 = 155* or *140 + 10 = 150, ES = 155*

– Line 7 : Case b => We have *155 + 25 = 180* or *140 + 20 = 160, ES = 180*

– Line 8 : Case a => *ES = 180 + 15 = 195*

COMMENT 5.3.– There is no reason for the values of ES to get continually higher.

⇒	1	2	3	4	5	6	7	8	*ES*
1		120							0
2			20	15	20				120
3						10	20		140
4					0				135
5						15			140
6							25		155
7								15	180
8									195
LF									

Table 5.8. *Double-entry matrix containing the earliest dates*

5.8.4. *Latest finish dates*

The algorithm for determining these dates is as follows:

– we copy the last value calculated for *ES* into the last column in line *LF*;

– we look for the value in the same column, just above the diagonal, from which we subtract the value of the last column of *LF* and place it in the column to the left;

– remaining in this column, we look for the value just above the diagonal. Three cases are possible:

- the value is alone on the line,

- several values exist on the line,

- the value does not exist, but there are other values on the line;

– in the first case, we subtract it from the previously found value;

– in the second and third cases we subtract all the values from the previously found values column by column and choose the lowest. Careful: an inexistent value is never considered to be equal to *0*. When there is no value, no calculation is possible;

– we continue in this way until we reach the first column.

Using this algorithm in our example gives the following calculations:

Column 8: *LF = 195*

Column 7: *LF = 195 – 15 = 180*

Column 6: Case a => *LF = 180 – 25 = 155*

Column 5: Case a => *LF = 155 – 15 = 140*

Column 4: Case a => *LF = 140 – 0 = 155*

Column 3: Case c => We have *155 – 10* or *180 – 20, LF = 145* (the smallest difference)

Column 2: Case b => We have *145 – 20, 140 - 15* or *140 – 20, LF = 120*

Column 1: Case a => *LF = 120 – 120 = 0*

⇒	1	2	3	4	5	6	7	8	*ES*
1		120							0
2			20	15	20				120
3						10	20		140
4					0				135
5						15			140
6							25		155
7								15	180
8									195
LF	0	120	145	140	140	155	180	195	

Table 5.9. *Double entry matrix presenting the earliest dates and latest dates*

COMMENT 5.4.– There is no reason for the values of LF to get continually lower.

The values of *ES* in line 1 and of *LF* in column 1 must always be equal (in our example *0*) otherwise one or more of the calculations are incorrect.

5.8.5. *Critical path*

When the table is finished, we determine the critical path. To do so, we go to the line of *LF* column 1 and we compare its value with the diagonally opposite value in column *ES*. We then continue in the same way until the final column. In the event they are the same, we check the number in the column.

Using this method on the table of our example gives:

LF column 1 = *ES* line 1 => No. 1 checked

LF column 2 = *ES* line 2 => No. 2 checked

LF column 3 ≠ *ES* line 3

LF column 4 ≠ *ES* line 4

LF column 5 = *ES* line 5 => No. 5 checked

LF column 6 = *ES* line 6 => No. 6 checked

LF column 7 = *ES* line 7 => No. 7 checked

LF column 8 = *ES* line 8 => No. 8 checked

⇒	•	♦	3	4	◖	◗	⬤	⬇	ES
1		120							0
2			20	15	20				120
3						10	20		140
4					0				135
5						15			140
6							25		155
7								15	180
8									195
LF	0	120	145	140	140	155	180	195	

Table 5.10. *The double-entry matrix and determining the critical path*

This means that the critical path passes through the vertices 1, 2, 5, 6, 7 and 8. When this path is drawn on the PERT graph, the associated critical paths can be found: A, D, F, G and I.

The overall duration of the project is *195 min* (*ES* of the first line or *LF* of the last column).

5.9. Methodology for planning with MPM

5.9.1. *A history*

At the same time as the development of the PERT and CPM methods in the USA, in France, Bernard Roy, in 1958, created the *Méthode des potentiels Métra* (MPM) to build the "France" ocean liner and then power plants for EDF[6].

It is very widely used and project management software often uses graphs based on its formalism.

It was designed to be more flexible than the PERT method, but if the author had to give an opinion, he would say that it is first and foremost a question of habit. Nevertheless, it is true that its graph is perhaps easier to build and read.

5.9.2. *Formulating the graph*

The MPM method is based on a graph in which each summit is a task and each edge is a succession constraint.

It uses two symbols:

– vertices which can be represented in different forms: rectangles, squares or circles. They are divided into several parts which can be attributed the name of the task, its duration and its earliest and latest dates, and even the free slack and total slack. When the duration is not presented within the vertex, it is placed on the edge(s) following it;

– edges which are a simple arrows. They indicate a value specifying a duration when it is not defined within the vertex.

---

6 Electricité de France, primary producer and supplier of electricity in France.

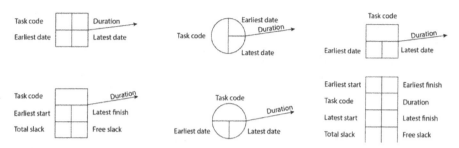

Figure 5.18. *Several possible representations of a task and of an associated constraint (edge) within an MPM graph*

As with the PERT method, we start with precedence table.

To draw an MPM graph, we start by placing an initial task with a duration *0*, which we call "Start" and corresponds to the start of the project (or the start of the subproject, if the graph being drawn is part of a larger project).

There is always an obligatory task called "Finish", which closes the project.

In an MPM graph, time passes from the left to the right of the graph. The final task is always further right than the start task.

Starting with the "Start" task, the tasks will be interlinked with edges that respect the anteriority constraints.

5.9.3. *Building rules*

Tasks can be assembled in several ways, thus creating dependency rules:

– sequential or successive tasks: a task cannot start before another is finished. There is a successor task and a predecessor task;

– simultaneous or parallel tasks: two or more tasks start at the same time;

– convergent tasks: two or more tasks meet at the same task.

The dependencies created in the MPM are the same as those in PERT (see section 5.5.1).

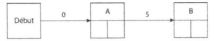

Start, A and B are sequential or successive

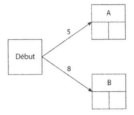

A and B are simultaneous or parallel and come from start

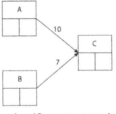

A and B converge towards C

Figure 5.19. *The different possible ways of assembling tasks in an MPM graph*

To create a graph, certain rules must be respected while it is being built so it remains clear and comprehensible:

– place tasks from left to right creating columns that will form different levels and make it easier to read the graph;

– do not cross edges where possible;

– draw edges as a straight line;

– avoid having edges that are too long;

– draw edges whose angles are as big as possible in relation to the edges.

5.9.4. *Earliest and latest dates*

Earliest dates are obtained by traversing the graph from left to right and successively adding up (at each task) the value of the earliest date at their duration, placed on the edge(s) following it (or within the vertex of the task in certain types of representations). In the case of convergence, we choose the largest value. The earliest date of the task following succeeding the start date is generally *0*, unless in the case of a subproject depending on a more substantial project.

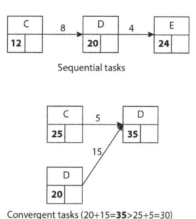

Figure 5.20. *The two ways of calculating an earliest date on an MPM graph*

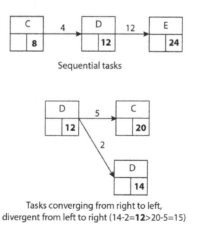

Figure 5.21. *The two ways of calculating a latest date on an MPM graph*

To get the latest dates, we traverse the graph from right to left, from the project finish task until the start task, successively subtracting the duration of each task, indicated by the edge(s) preceding the given task (or within the vertex of the previous tasks on certain types of representations). In the case of convergence, we choose the difference with the smallest value.

5.9.5. *Determining the critical path*

Critical tasks are those whose earliest and latest dates are the same. They therefore form a route called the critical path. There can be several possible routes when graphs become complex.

As for the PERT method, any modification to the duration of a critical task will have consequences on the overall duration and the finish date of a project. However, a change to the duration of one or more non-critical tasks will not necessarily have an impact and work may be able to be finished within the predicted timeframe if the available slacks are respected (see above section 5.6). Exceeding the slacks will change the critical path and the tasks composing it.

5.10. Example of determining a critical path with MPM

We will now return to the scenario previously described (see section 5.5) for the PERT graph and apply it to MPM.

As the precedence table has not changed, we will keep it and build our graphs from it.

COMMENT 5.5.– I have chosen one of the most common modes of representation for the example we are going to discuss i.e. a square containing the task code, its earliest date and latest date. The duration is indicated by the edge following the task. Where necessary, I will specify any changes for other modes of representation.

5.10.1. *Creating the graph*

Create the start task, input its name and then do the same with the successors, inputting their code while respecting the predecessors until the final task. On each edge following the task (or on the task itself depending on the type of representation) you have just entered, place its duration. Do not forget to finish the graph by adding a finish task, as you advance align tasks vertically one above the other.

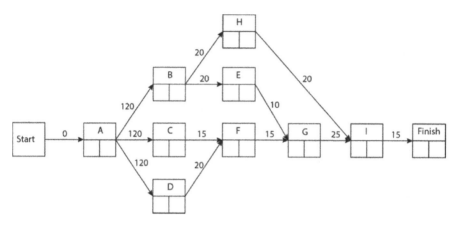

Figure 5.22. *MPM graph with its tasks and their duration is placed along each edge*

5.10.2. *Determining the earliest date of each task*

The approach to calculate the earliest dates for our example is detailed below:

COMMENT 5.6.– To be as clear as possible, a letter is considered to be a task.

– place *0* as the duration of the "Start" task;

– A has B, C and D as successors. Note in each of these the total of *0 + 120*, that is *120*;

– from B we can reach H and E. Add the earliest date of B with its duration, that is *120 + 20*, and put *140* in H and E;

– from C and D we can reach F as the two tasks converge. We will therefore add *120 + 15* and *120 + 20* and put the largest value, i.e. *140* in F;

– from E and F we can reach G as two tasks also converge here. We add *140 + 10* and *140 + 15* and choose the largest value, i.e. *155*, which we place in G;

– from H and G we can reach I, two tasks converge here again. We calculate the totals of *140 + 20* and *155 + 25* and take the largest, i.e. *180*, which we place in I;

– to finish we put the sum of *180 + 15*, i.e. *195* in "Finish".

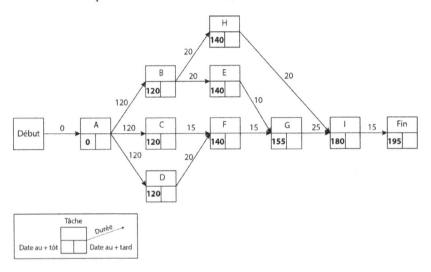

Figure 5.23. *The MPM graph of our example with the earliest dates calculated*

5.10.3. *Determining the latest dates of each task*

The approach to calculate the latest dates for our example is detailed below:

– we go to "Finish" and give it the value of its start date, i.e. *195* as its latest date;

– to go from Finish to I, we calculate *195 – 15*, i.e. *180*, which becomes the latest date of I;

– from I, we can return to G and H, *180 – 25*, i.e. *160* for H and *180 – 25*, i.e. *155* for G;

– from G, we can return to E and F, *155 – 10*, i.e. *145* for E and *155 – 15*, i.e. *140* for F;

– B can be reached via E or H which converge. We get *145 – 20* and *160 – 20* respectively. We attribute *125* to B;

– from F we can return to C, *140 – 15*, i.e. *125*;

– from F, we can also return to D, *140 – 20*, i.e. *120*;

– B, C and D converge toward A. We get *120 – 120, 125 – 120, 120 -120,* respectively, i.e. *0* as a latest date for A.

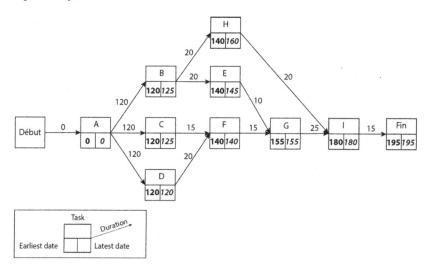

Figure 5.24. *MPM graph of our example with the earliest and latest dates calculated*

5.10.4. *Determining the critical path(s)*

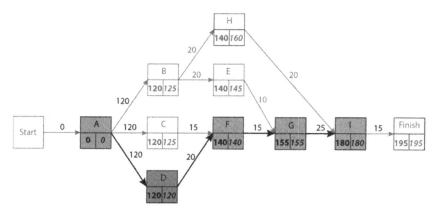

Figure 5.25. *The critical tasks (A, D, F, G and I) in the MPM graph of our example*

In our example, that critical tasks are A, D, F, G and I. Note that the result is identical to that found by the PERT method, which is no surprise.

5.10.5. *Slacks*

As for PERT, from the MPM graph we can calculate the free, independent and total slacks of each task.

Nevertheless, determining the latest finish date is less evident in the case of divergent edges from the same task.

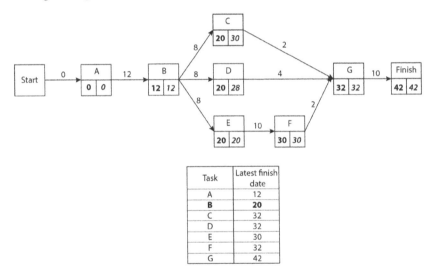

Task	Latest finish date
A	12
B	**20**
C	32
D	32
E	30
F	32
G	42

Figure 5.26. *Determining the latest finish on an MPM graph. Here there is the particularity of task B that converges toward C, D and E*

In this situation, we apply the following rule: the latest finish date of the task in question is the lowest earliest finish date of all the divergent tasks that follow it.

5.11. Probabilistic PERT/CPM/MPM

During the development of a project, there are always uncertainties relating to one or more tasks. Implementing a probability calculation will define a project duration attached to a probability.

Each task is a variable that only rarely has a constant duration that can be defined in advance with any certainty. At best, we can establish an estimation with varying degrees of probability of its duration.

If we apply this principle, the duration of the path traversing the graph, which is a total of values, also becomes a probable result.

This duration obeys a law of distribution similar to the Laplace–Gauss law. It is represented as a bell curve.

To be consistent, probabilistic task planning must respect certain conditions:

– weak distribution in the order of magnitude of task duration;

– strong independence between the task durations of the project;

– a high number of tasks resulting in paths with more than a half-dozen tasks[7].

It is worth saying that an uncertain task duration does not mean that it can take any value, instead it should take consistent values known with greater or lesser degrees of accuracy.

Project planning with a probabilistic method dates back to 1962. It was developed by C.E. Clark.

The three main objectives of the probabilistic method are to provide an average project duration, to determine a probability for finishing within a chosen duration and to define its duration according to a given probability.

5.11.1. *Probability of the tasks*

The duration of each task is distributed according to three values: minimum or optimistic, maximum or pessimistic and expected, more probable, likely or realistic. This probability distribution follows the Bêta law (also known as the "PERT law" in project management).

---

7 Experts disagree on this number, but they all agree that the minimum acceptable amount is 4 tasks on a path.

It is formulated as follows:

$$t_m(T_i) = \frac{t_o(T_i) + 4t_v(T_i) + t_p(T_i)}{6}$$

where

t_m : Average duration (or Estimate of expected activity time)

t_o : Optimistic duration

t_v : Likely duration

t_p : Pessimistic duration

T_i : Task in question

Note that this formula considers the probable duration to be four times more likely to occur than the optimistic or pessimistic duration. Though rarely the case, differences with reality will be minimized or increased, thus providing a small difference for a high number of tasks.

We also need to calculate the variance of each task:

$$Var(T_i) = \left(\frac{t_p(T_i) - t_o(T_i)}{6} \right)^2$$

from which we can calculate the standard deviation:

$$\sigma_i = \sqrt{Var(T_i)} \text{ or } \sigma_i = \frac{t_p(T_i) - t_o(T_i)}{6}$$

As mentioned above, we consider that the sum of the probable durations of tasks tends to follow a normal law with which we can define the average duration of the project as being:

$$T_{Project} = \sum_{i=1}^{n} t_{Pr}(T_i)$$

Its standard deviation will be:

$$\sigma_{Project} = \sqrt{\sum_{i=1}^{n} \sigma_i^2}$$

5.11.2. *Its implementation on an example*

Let us pick up once again the example used in section 5.5.1. We modify the precedence table by specifying the three durations (pessimistic, probable and optimistic), which are duly evaluated, for each of the tasks according to our experience or observations made on the ground.

Task	Description	Pessimistic duration (min)	Likely duration (min)	Optimistic duration (min)	Predecessors
A	Transporting the material	140	120	58	–
B	Unloading front and back loudspeakers systems	30	20	15	A
C	Unloading and installing the amplification system and cables	25	15	12	A
D	Unloading the mixing console	28	20	17	A
E	Wiring front speakers	20	10	8	B
F	Installing cabling console and amplification system	25	15	12	C; D
G	Assembling, putting together and connecting front speakers	40	25	20	E; F
H	Wiring back speakers	32	20	15	B
I	Test installation	25	15	10	G; H

Table 5.11. *The example from section 5.5.1 with new durations*

We then determine the critical path with the method of our choice –
PERT/CPM or MPM – to obtain the critical tasks that we add to the table to
finalize our probabilistic calculation.

5.11.3. Calculating average durations and variance

We know that the critical tasks in our example are A, D, F, G and I,
giving us the table below which we use to calculate their average durations
and variances.

T_i	$t_p(T_i)$	$t_v(T_i)$	$t_o(T_i)$	$t_m(T_i)$	$Var(T_i)$
A	140	120	58	113.000	186.778
D	28	20	17	20.833	3.361
F	25	15	12	16.167	4.694
G	40	25	20	26.667	11.111
I	25	15	10	15.833	6.250

Table 5.12. *Average durations and variances*

5.11.4. Calculating the average duration of the project

The average duration is equal to the sum of column $t_m(T_i)$, i.e.:

$T_{Project} = 192.5\ min$

5.11.5. Calculating the probability of finishing the project according to a chosen duration

To determine the probability that this project will finish in *215 min*, we
use the table of the standard normal distribution (see Appendix 2) once we
have calculated the value of the relative variable change *i*, i.e.:

$$i = \frac{T_x - T_{Project}}{\sigma_{Project}}$$

where:

T_x: Chosen duration

$T_{project}$: Average duration of the project

$\sigma_{Project}$: Standard deviation of the project

We calculate the variance, which is equal to the sum of the column $Var(Ti)$, i.e.:

$$Var_{Project} = 212,194$$

Then the standard deviation:

$$\sigma_{Project} = \sqrt{Var_{Project}} = \sqrt{212,194} \; ; 14,567$$

To obtain i :

$$i = \frac{215 - 192,5}{14,567} \approx 1,54$$

By putting i in the table of the standard normal distribution, we obtain a probability of *93.82%* that the project will have a duration of *215 min*.

5.11.6. *Calculating the duration of the project for a given probability*

Here we look to determine the duration of the project T with a probability of *95%*.

In the table of the standard normal distribution, we find that the value i is *1.64* for *94.95%* and *1.65* for *95.05%*.

To find out a value close to i for a probability of *95%*, we calculate the linear interpolation i.e.:

$$i_{90} = \frac{1.64(95.05 - 95) + 1.65(95 - 94.95)}{(95.05 - 95) + (95 - 94.95)} = 1.645$$

We can now calculate the duration of the project:

$$T_{p95} = T_{Project} + (i_{95} \times \sigma_{Project})$$

$$T_{p95} = 192.5 + (1.645 \times 14.567) \approx 216.462\text{min}$$

5.12. Gantt chart

We owe the Gantt chart to its inventor Henry Laurence Gantt[8] who created it in 1910.

This tool models the planning of tasks in a project in the form of a connected, oriented and valued graph which graphically shows the progress of a project.

It has two perpendicular axes. The horizontal axis of the abscissa represents time and the vertical axis of the ordinates represents the tasks in the project.

Its main benefit is that it is easy to read: in one glance, we can see the delay or the progress of the project. It enables the dates for performing the tasks in a project and existing slacks to be determined with ease.

It is easy to create a Gantt chart from the slacks table, which shows all the required elements.

5.12.1. *Creating the chart*

To build a Gantt chart, we follow the process detailed below:

– we draw the frame of reference with two axes. The abscissa axis, time, will be standardized to cover the maximum length of the project, divided into a schedule or time (months, weeks, days, hours, minutes, etc.). The ordinate axis will be broken down into tasks, the most recent being placed at its upper end;

---

8 1861–1919, American mechanic who worked with Frederick Winslow Taylor. He is attributed with the discovery of his famous chart, though a Polish engineer Karol Adamiecki first developed it in 1896.

– to plot the different elements in the chart, we create a key that will take into account the critical and non-critical tasks, the free slacks and the total slacks;

– we then draw the different tasks in the form of horizontal bars whose lengths correspond to their duration. The bars representing critical and non-critical tasks are different and respect the key. The outline of the bar extends to the earliest start date to its latest finish date;

– we continue with the total and free slacks which are placed at the far right end of each task. If a task has two types of slack, they are superimposed;

– finally, we can close the chart by adding arrows between the tasks indicating their succession constraints.

COMMENT 5.7.– Fictive tasks do not exist in the Gantt chart. For greater comprehension, we can add the duration of each of the tasks at the center of each bar.

5.12.2. Example

To implement the procedural steps described above, we will return once again to the example used for the PERT and MPM graphs (see section 5.5).

The slacks table from section 5.6 provides all the elements we need.

Task	Start step	Finish step	Duration	Earliest start	Latest start	Earliest finish	Latest finish	Total slack	Free slack
A	1	2	120	0	0	120	120	0	0
B	2	3	20	120	125	140	145	5	0
C	2	4	15	120	125	140	140	5	5
D	2	5	20	120	120	140	140	0	0
E	3	6	10	140	145	150	155	5	5
F	5	6	15	140	140	155	155	0	0
G	6	7	25	155	155	180	180	0	0
H	3	7	20	140	160	160	180	20	20
I	7	8	15	180	180	195	195	0	0
X	4	5	0	135	140	135	140	-	-

Table 5.13. *All the data required for the Gantt chart*

5.12.2.1. *Drawing the frame of reference*

The two axes drawn and standardized. As the duration of the project does not exceed *195* min, the abscissa axis is regulated until *210* min. The ordinate axis presents 9 necessary tasks from A to I. We added a key which will apply to drawing the task bars.

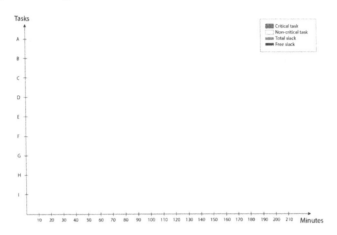

Figure 5.27. *Frame of reference for the Gantt chart for our example*

5.12.2.2. *Drawing the tasks*

On each of the bars, we added the duration for greater comprehension.

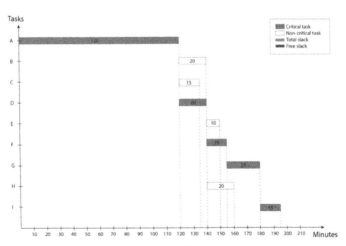

Figure 5.28. *Gantt chart and its tasks*

5.12.2.3. *Adding the slacks*

The total and free slacks are added to the end of each task that has them. Note that they are superimposed for C, E and H.

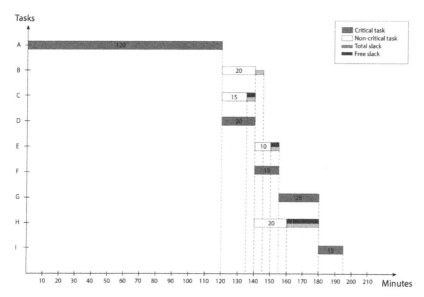

Figure 5.29. *The chart with its slacks*

5.12.2.4. *Adding constraint arrows*

To make the constraints clearer, we can add arrows showing the chain:

– task A is followed by B, C and D;

– task B is followed by E and H;

– tasks C and D are followed by F;

– tasks E and F are followed by G;

– tasks G and H are followed by I.

5.13. PERT-MPM costs

Until this point we have not mentioned costs, though they have a strong connection with time management. It is therefore possible to define a cost-duration relation.

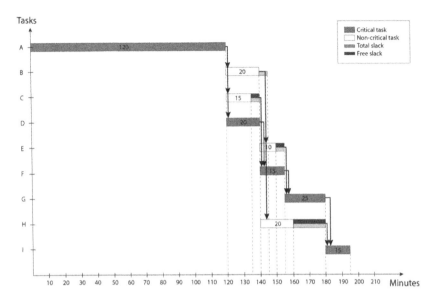

Figure 5.30. *Arrows showing the succession constraints*

5.13.1. *Method*

For each task in the project we define:

–a *normal cost C* which corresponds to the lowest cost, i.e. with a minimum amount of resources to meet the required work;

– a *likely cost T* which is consistent with the normal cost;

– an *accelerated cost* C_a, which corresponds to the minimum time required to perform the task with sufficient resources;

– an *accelerated duration* T_a which corresponds to the minimum conceivable duration for the task.

We can see that there is a linear relationship between time and cost such that we can define a *marginal acceleration cost* C_m which will show the additional cost that a task will require to be performed within its reduced duration.

$$C_m = \frac{C_a - C}{T - T_a}$$

To apply a PERT/MPM cost method, we need to analyze the graph to seek to reduce the duration of the project with a view to obtaining a minimum cost.

As the duration of the project is by definition linked to the critical path, our effort to optimize cost will affect the critical tasks.

A number of rules can be applied enabling a solution to be obtained that is not necessarily optimal but which is satisfactory:

– out of the critical tasks, reduce those which have the lowest marginal acceleration cost;

– then reduce the duration of the successive critical tasks.

5.13.2. *Example*

Let us return to our favorite scenario, for which we have the following duration/cost table (only the critical tasks are taken into consideration):

Task	Duration – T (min)	Predecessor	Maximum reduction (min)	Cost – C ($/h)	Accelerated cost – C_a ($/h)
A	120	–	62	55.00	85.00
B	20	A	5		
C	15	A	3		
D	20	A	3	24.00	28.70
E	10	B	2		
F	15	C; D	3	24.50	29.00
G	25	E; F	5	27.00	32.50
H	20	B	5		
I	15	G; H	5	28.00	36.00

Table 5.14. *Durations, costs and accelerated costs*

We calculate the marginal acceleration costs C_m once we have transformed all the data into hours and take the minimum durations T_a from the previous table.

Task	Duration – T (min)*	Minimum duration – T_a (min)*	Cost – C ($/h)	Accelerated cost – C_a ($/h)	Marginal acceleration cost – C_m ($/h)
A	120 (2h)	58 (0.967h)	55.00	85.00	29.04
D	20 (0.333h)	17 (0.283h)	24.00	28.70	94.00
F	15 (0.250h)	12 (0.200h)	24.50	29.00	90.00
G	25 (0.417h)	20 (0.333h)	27.00	32.50	65.48
I	15 (0.250h)	10 (0.167h)	28.00	36.00	96.39

Table 5.15. *Marginal costs*

The values of duration in hours have been rounded up.

We apply the rules described above:

– the critical task with the lowest marginal acceleration cost is task A (*29.04 $/h*), let us reduce it by *62 min* (T_a = *58 min*). The critical path and the slacks remain the same. The project is reduced to *133 min*;

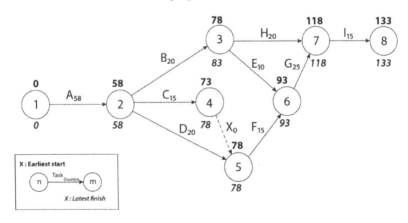

Figure 5.31. *The project once the duration of task A has been reduced to 58 min*

– the successive task that has the lowest marginal cost is G (*65.48 $/h*), which we can reduce by *5 min* (T_a = *20 min*). We apply this reduction. The critical path remains the same and the project is reduced to *128 min*;

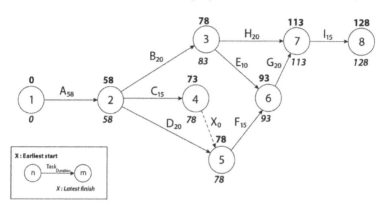

Figure 5.32. *The project once the duration of task G has been reduced to 20 min*

– let us now go on to F (*90.00 $/h*). We can reduce it by *3 min* (T_a = *12 min*). The critical path remains the same: the project now lasts *125 min*;

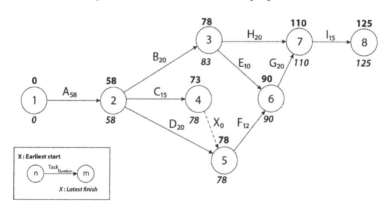

Figure 5.33. *The project once the duration of task F has been reduced to 12 min*

– the next task is D (*94.00 $/h*), but we cannot reduce it to its minimum value, T_a = *17 min*, as it would change the critical path. We will therefore limit it to a reduction of *2 min*, resulting in a reduced duration to *18 min*. The duration of the project is therefore *123 min*.

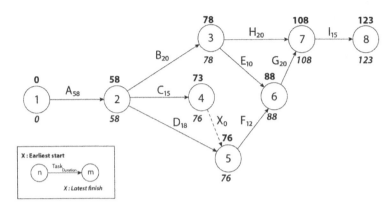

Figure 5.34. *The project once the duration of task D has been reduced to 18 min*

In terms of costs, and only considering those linked to the critical tasks (as the duration of the non-critical tasks has not been modified, their costs remain the same), we can calculate the overview table below:

Task	T (h)	Reduced duration (h)	C ($/h)	Reduction (h)	C_m ($/h)	Add. cost ($/h)	Normal cost ($)	Optimized cost ($)
A	2.000 (120min)	0.967 (58min)	55.00	1.033	29.04	30.00	110.00	83.19
D	0.333 (20min)	0.300 (18min)	24.00	0.033	94.00	3.10	7.99	10.30
F	0.250 (15min)	0.200 (12min)	24.50	0.050	90.00	4.50	6.13	9.40
G	0.417 (25min)	0.333 (20min)	27.00	0.084	65.48	5.50	11.26	14.49
I	*0.250 (15min)*	–	*28.00*	–	–	–	*7.00*	*7.00*
Total							142.38	124.38

Table 5.16. *Cost overview*

where:

– reduced duration: minimum acceptable reduction without changing the critical path;

– *reduction = T – Reduced duration;*

– *add. cost = Reduction $\times C_m$;*

– *normal cost = T $\times C$;*

– *optimized cost = Reduced duration $\times C$ + Add. cost.*

COMMENTS 5.8.–

All durations are expressed in hours and have been rounded up in this table.

The "Reduced duration" corresponds to the values listed during the application of the rules linked to the PERT/MPM cost.

The "Add. cost" is the additional cost due to the duration of the task being reduced.

The "Optimized cost" is the cost of the task once its duration has been reduced.

In conclusion, it can be said that it is more economic (*$124.38*) to reduce the duration of the project to *123 min* even though it implies additional marginal costs for certain tasks.

5.14. Exercises

5.14.1. *Exercise 1*

Let us consider the precedence table below.

Task	Duration (day)	Predecessor
A	5	–
B	8	C
C	3	–
D	7	A
E	10.5	C
F	8	B, D
G	2	E, F

Table 5.17. *Table of predecessors*

Questions:

1) Knowing that the project will start on the 20th May of this year and that the Saturday and Sunday are non-working days, use the PERT method to determine the duration of the project, its finish date and the critical tasks and path.

2) Create the table of the slacks and calculate the total slacks and free slacks.

3) Create the Gantt chart.

5.14.2. Exercise 2

Let us look at the PERT graph below (durations are in days):

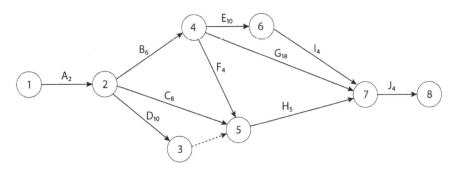

Figure 5.35. *The PERT graph*

Questions:

1) Create the precedence table.

2) Determine the critical tasks and the duration of the project with a double entry table.

3) Calculate the total and free slacks for the non-critical tasks.

5.14.3. Exercise 3

Let us look at the project represented by the precedence table below:

Task	Optimistic duration (day)	Likely duration (day)	Pessimistic duration (day)	Predecessor
A	1	2	4	–
B	4.5	6	8	A
C	16	20	25	B
D	9	10	14	B
E	2	3	4	D, F
F	1.5	2	4	C
G	2	4	5.5	F
H	4	6	8	F
I	27	30	36	E, G, H
J	7	8	9	I
K	1	2	2.5	J

Table 5.18. *Durations and predecessors*

Questions:

1) Create the MPM diagram of this project, determine the critical tasks and its likely duration.

2) Calculate the average duration of the project.

3) What is the probability of this project finishing in 81 days?

4) What will be the duration of the project for a duration of 98%?

5.14.4. *Exercise 4*

Let us consider the project defined by the precedence matrix and the cost-duration table below:

	A	B	C	D	E	F	G	H	I
A		X	X	X					
B						X			
C					X	X			
D							X	X	
E							X	X	
F							X	X	
G									X
H									X
I									

Table 5.19. *The precedences for exercise 4*

Questions:

1) Determine the duration of the project and the critical tasks by the PERT method.

2) Calculate the normal cost of the project.

3) Calculate the optimized cost of the project and its duration.

Task	Duration – T (j)	Maximum reduction (j)	Cost – C (€/j)	Accelerated cost – C_a (€/j)
A	120	25	300.00	385.00
B	180	40	420.00	510.00
C	3	0.5	330.00	490.00
D	250	30	240.00	360.00
E	60	8	300.00	500.00
F	90	21	600.00	790.00
G	240	45	320.00	420.00
H	180	50	270.00	500.00
I	30	6	480.00	560.00

Table 5.20. *Costs and durations for exercise 4*

5.14.5. *Solution to exercise 1*

Question 1:

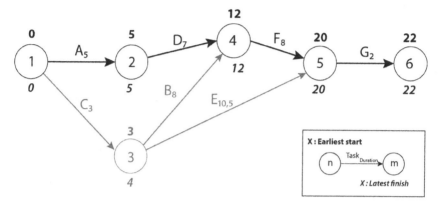

Figure 5.36. *The PERT diagram*

The duration of the project is *22 days*.

Its finish date, taking into account the four weekends and the start date of the project planned for the 20th May will be: *22 + (4 x 2) = 30 days* after the 20th May included, which makes it the evening of the 18th June.

The critical path passes through vertices 1, 2, 4, 5 and 6.

The critical tasks are A, D, F and G.

Question 2:

Task	Start step	Finish step	Duration (day)	Earliest start	Earliest finish	Latest finish	Total slack	Free slack
A	1	2	5	0	5+0=5	5	5–0–5=0	5-5=0
B	3	4	8	3	8+3=11	12	12–3–8=1	12-11=1
C	1	3	3	0	3+0=3	4	4–0–3=1	3-3=0
D	2	4	7	5	7+5=12	12	12–5–7=0	12-12=0
E	3	5	10.5	3	10.5+3=13.5	20	20–3–10.5=6.5	20-13.5=6.5
F	4	5	8	12	8+12=20	20	20–12–8=0	20-20=0
G	5	6	2	20	2+20=22	22	22–20–2=0	0

Table 5.21. *Total and free slacks*

Question 3:

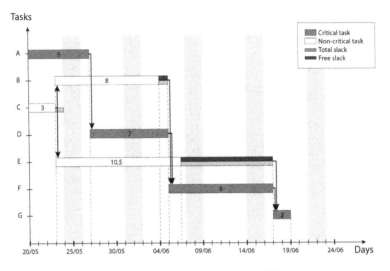

Figure 5.37. *Gantt chart with its key, all the tasks and slacks as well as the anteriority constraints*

5.14.6. *Solution to exercise 2*

Question 1:

Task	Duration (day)	Predecessor
A	2	–
B	6	A
C	8	A
D	10	A
E	10	B
F	4	B
G	18	B
H	5	C, D, F
I	4	E
J	4	G , H, I

Table 5.22. *Precedence table*

Question 2:

	❶	❷	3	❹	5	6	❼	⑧	i
1		2							0
2			10	6	8				0+2=2
3					0				2+10=12
4					4	10	18		2+6=8
5							5		8+4=12
6							4		8+10=18
7								4	18+8=26
8									26+4=30
j	2-2=0	8-6=2	21-0=21	26-18=8	26-5=21	26-4=22	30-4=26	30	

Table 5.23. *Double-entry matrix with its earliest and latest dates*

The critical path passes through the vertices 1, 2, 4, 7 and 8 meaning that the critical tasks are: A, B, G and J.

The duration of the project is *30 days*.

Question 3:

Task	Start step	Finish step	Duration (day)	Earliest start	Earliest finish	Latest finish	Total slack	Free slack
C	2	5	8	2	8+2=10	21	21-2-8=11	12-10=2
D	2	3	10	2	10+2=12	21	21-2-10=9	12-12=0
E	4	6	10	8	10+8=18	22	22-8-10=4	18-18=0
F	4	5	4	8	4+8=12	21	21-8-4=9	12-12=0
H	5	7	5	12	5+12=17	26	26-12-5=9	26-17=9
I	6	7	4	18	4+18=22	26	26-18-4=4	26-22=4

Table 5.24. *Total and free slacks*

5.14.7. *Solution to exercise 3*

Question 1:

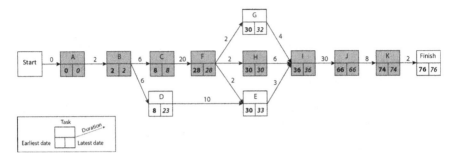

Figure 5.38. *The MPM diagram with its critical tasks*

The critical tasks are: A, B, C, F, H, I, J and K.

The likely duration of the project is *76 days*.

Question 2:

T_i	$t_p(T_i)$	$t_v(T_i)$	$t_o(T_i)$	$t_m(T_i)$	$Var(T_i)$
A	4	2	1	2.167	0.250
B	8	6	4.5	6.083	0.340
C	25	20	16	20.167	2.250
F	4	2	1.5	2.250	0.174
H	8	6	4	6.000	0.444
I	36	30	27	30.500	2.250
J	9	8	7	8.000	0.111
K	2.5	2	1	1.917	0.063

Table 5.25. *Calculation of the average durations
and the variances for the critical tasks*

The average duration of the project is *77.083 days* (sum of the column *tm(Ti)*)

Question 3:

$\sigma Project = 2.425$

$i = 1.61$

The probability of finishing the project in *81 days* is *94.63%*.

Question 4:

$i = 97.98\%$ for *2.05*

$i = 98.03\%$ for *2.06*

After interpolation, we get: $i = 98 = 2.054$

For a probability of *98%*, the project will last approximately *82.064 days*.

5.14.8. *Solution to exercise 4*

Question 1:

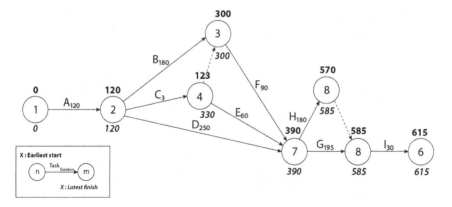

Figure 5.39. *The PERT graph*

The project will last *660* days and the critical tasks are A, B, F, G and I

Question 2:

Task	T(j)	C (\$/j)	Cost
A	120	300.00	36,000.00
B	180	420.00	75,600.00
C	3	330.00	990.00
D	250	240.00	60,000.00
E	60	300.00	18,000.00
F	90	600.00	54,000.00
G	240	320.00	76,800.00
H	180	270.00	48,600.00
I	30	480.00	14,400.00
Total			384,390.00

Table 5.26. *Normal cost calculation*

The normal cost of the project will be *\$384,390.00*

Question 3:

Task	T (j)	Maximum reduction (j)	T_a (j)	C (\$/j)	C_a (\$/h)	C_m (\$/j)
A	120	25	95	300.00	385.00	3.40
B	180	40	140	420.00	510.00	2.25
F	90	21	5	600.00	790.00	9.05
G	240	45	195	320.00	420.00	2.22
I	30	6	24	480.00	560.00	13.33

Table 5.27. *Accelerated and marginal costs calculation*

Reduction of G : *240 – 45 = 195 days*. The duration of the project is *615 days*.

Reduction of B : *180 – 20 = 160 days* (after 20 days, the critical path changes). The duration of the project is *595 days*.

Task	T (j)	Reduced duration (j)	C ($/j)	Reduction (j)	C_m ($/j)	Add. cost ($/h)	Optimized cost ($)
A	120		300.00		3,40		36 000.00
B	180	160	420.00	20	2,25	45.00	67 245.00
C	3		330.00				990.00
D	250		240.00				60 000.00
E	60		300.00				18 000.00
F	90		600.00		9,05		54 000.00
G	240	195	320.00	45	2,22	99.90	62 499.90
H	180		270.00				48 600.00
I	30		480.00		13,33		14 400.00
Total							361 734.90

Table 5.28. *Optimized cost calculation*

The optimized cost of the project will be *$361,734.90.*

Maximum Flow in a Network

6.1. Maximum flow

A transport network is a graph without loops. Each edge in the graph is associated with a number c, capacity, with $c \geq 0$.

This network must confirm the two following hypotheses:

1) Just one of its vertices (nodes) has no antecedents (predecessors), all the others have at least one. This vertex is the point of access to the network, it is often called the **source** s.

2) Just one of its vertices (nodes) has no successors, all the others have at least one. This vertex is the point of exit of the network, it is often called the **sink** t.

The flow f carried by the network is a function associating a quantity with each of the edges a. This function $f(a)$ represents the value of the flow carried by the edge. It comes from the source and moves toward the sink.

One of the fundamental rules of a flow is that the sum of the values carried by each of the edges entering a vertex (other than s and t) must be equal to the sum of the values carried by the edges exiting this same vertex. In sum, the total quantity of the flow entering a vertex is equal to the total exiting quantity.

Digraphs confirm **node law** or **Kirchhoff**[1] **law**: the flow is constant and does not form a queue in a vertex. This is the principle of flow conservation.

A flow traversing an edge must never exceed its capacity. For every edge *a*, we have: $0 \leq f(a) \leq c(a)$. This flow is therefore said to be **compatible** or **admissible**.

Moreover, a flow is said to be **complete** if for each path going from the source to the sink there is at least one **saturated edge**, i.e. if the flow traversing it is equal to its capacity.

A **saturated path** is a path with at least one saturated edge.

There are many applications relating to flow and network problems and they concern many domains, including:

1) logistics and transporting merchandise by road, rail, boat, etc;

2) energy with EDF networks, distribution from plants to clients, etc;

3) information media, whether it be the phone network, local networks, the Internet, etc;

4) fluid distribution with canal systems for running water, petrol industry with pipelines, etc.

One of the most commonly encountered and classic problems relating to networks is searching for a maximum flow.

Each edge in a graph has a maximum capacity which gives the highest value of the flow it is likely to transport. We then look to determine the flow with a maximum value at a given place.

Each of the flow units transported can also be associated with a cost on each of the edges, thus enabling a minimum cost for a maximum flow to be found.

To find solutions to these problems, we can use different algorithms, such as those developed by Ford and Fulkerson, Dinic, Dinic and Karzanov, Edmonds and Karp.

---

1 Gustav Robert Kirchhoff, 1824–1887, was a German physicist specializing in electrodynamics, the mathematical theory of elasticity and radiation physics.

This book will describe the first two algorithms, which are by far the most widely used.

6.2. Ford–Fulkerson algorithm

This algorithm is named after its authors[2] and was published in 1956. It is iterative and at each step looks for a flow that meets the capacity constraints and then tries to improve it.

To realize each of the iterations we use a **marking** technique using the symbols " + " or " - " to show whether there can be an improvement in the flow. We therefore create an **augmenting chain**.

From this marking, we can calculate the flow N of the augmenting chain following the following rules:

$N = min\{N^+, N\}$

$N^+ = min\{c(i, j) - f(i, j)\}$ where $(i, j) \in A^+$

$N^- = min\{f(i, j)\}$ where $(i, j) \in A^-$

where:

A^+ : total forward edges (direction: from the source toward the sink +) of the path carrying the augmenting chain;

A^-: total backward edges (direction: from the sink toward the source -) of the path carrying the augmenting chain;

$c(i, j)$: edge capacity (i, j);

$f(i, j)$: edge flow (i, j);

$N+$: value of the flows of the edges in forward direction;

$N-$: value of the flows of the edges in backwards direction;

N : total value of the flow of the augmenting chain.

---

2 Lester Randolph Ford, 1886–1967, and Delbert Ray Fulkerson, 1924–1976, American mathematicians.

When N is known, we can calculate the value of the flow carried by each of the edges of the augmenting chain with the following rules:

1) if the edge is in a forward direction then $f(i, j) = f(i, j) + N$;

2) if the edge is in a backward direction then $f(i, j) = f(i, j) - N$.

6.2.1. *Presentation of the algorithm*

The algorithm is based on three steps:

1) initialization, where each edge in the network is attributed a flow equal to 0 and the source is marked $+$;

2) marking-calculation, which looks for an augmenting chain using the marking method, then calculates the value of the flow N of this same chain;

3) updating the flow, which calculates the new flows that will be carried by each of the edges of the augmenting chain.

We consider the following variables:

s : source of the graph;

t : sink of the graph;

N : total value of the flow of the augmenting chain;

i : start vertex of an edge;

j : arrival vertex of an edge;

(i, j) : an edge;

c : capacity;

f : flow;

F : maximum flow.

The steps are presented below:

– Initialization
attribute 0 to F
attribute a flow f equal to 0 to all (i, j)
mark as $+$ the vertex s

– Marking-calculation
go to s
as long as j is not equal to t
    if i is marked + or - , j is not marked and $f(i,j) < c$ (unsaturated) then
        mark j with +
    end if
    if i is marked + or - , j is unmarked and $f(j, i) > 0$ then
        mark j with –
    otherwise
        stop, maximum flow
    end if
    calculate N
    calculate $F = F + N$
end as long as

– Updating the flow
go to s
as long as j is not equal to t
    if j is marked + then
        $f(i, j) = f(i, j) + N$
    if j is marked - then
        $f(i, j) = f(i, j) - N$
end as long as
remove marking except for s
start marking again

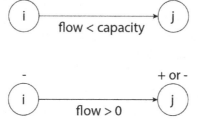

Figure 6.1. *The principle of marking*

6.2.2. *Application on an example*

Let us consider the graph below:

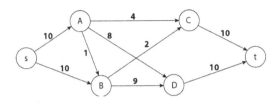

Figure 6.2. *The graph of our example*

We determine the maximum flow that can support this network by applying the Ford–Fulkerson algorithm.

Step 1: Initialization

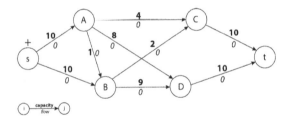

Figure 6.3. *Initialization*

The source *s* is marked + and the flows *f* of each of the edges are put at *0*. The maximum flow *F* is equal to *0*.

Step 2: Marking-calculation no. 1

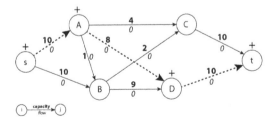

Figure 6.4. *The graph after marking no. 1 (the dotted lines are the augmenting chain)*

The marked chain is *sADt* and has a total flow value:

$$N^+ = \min\{10 - 0 \ ; \ 8 - 0 \ ; \ 10 - 0\} = 8 \text{ and } N^- = \varnothing \Rightarrow N = 8$$

COMMENT 6.1.– sADt was chosen because the capacity of the edges enables a total flow of 8 to be obtained. It must not be forgotten that we are looking to obtain the highest maximum flow possible. However, the choice of any other path would give the same result at the end of the algorithm.

$$F = 0 + 8 = 8$$

Step 3: Updating flow no. 1

The new flows of the edges of the augmenting chain are:

$$f(s, A) = 0 + 8 = 8 \ ; \ f(A, D) = 0 + 8 = 8 \ ; \ f(D, t) = 0 + 8 = 8$$

i.e. the graph in Figure 6.5.

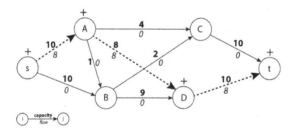

Figure 6.5. *The graph once the edges of the augmenting chain sADt have been calculated*

Step 2: Marking-calculation no. 2

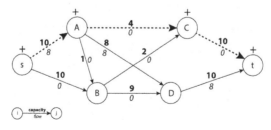

Figure 6.6. *The graph after marking no. 2 (the dotted lines are the augmenting chain)*

The marked chain is *sACt* and has a total flow value:

$$N^+ = \min\{10 - 8 ; 4 - 0 ; 10 - 0\} = 2 \text{ and } N^- = \varnothing \Rightarrow N = 2$$

Step 3: Updating the flow no. 2

The new flows of the edges of the augmenting chain are:

$$f(s, A) = 8 + 2 = 10 ; f(A, C) = 0 + 2 = 2 ; f(C, t) = 0 + 2 = 2$$

$$F = 8 + 2 = 10$$

i.e. the graph in Figure 6.7.

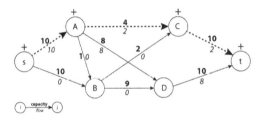

Figure 6.7. *The graph once the edges of the augmenting chain sACt have been calculated*

Step 2: Marking-calculation no. 3

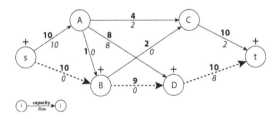

Figure 6.8. *The graph after marking no. 3 (the dotted lines are the augmenting chain)*

The marked chain is *sBDt* and has a total flow value:

$$N^+ = \min\{10 - 0 ; 9 - 0 ; 10 - 8\} = 2 \text{ and } N^- = \varnothing \Rightarrow N = 2$$

$$F = 10 + 2 = 12$$

Step 3: Updating the flow no. 3

The new flows of the edges of the augmenting chain are:

$$f(s, B) = 0 + 2 = 2 \; ; \; f(B, D) = 0 + 2 = 2 \; ; \; f(D, t) = 8 + 2 = 10$$

i.e. the graph in Figure 6.9.

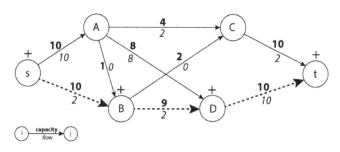

Figure 6.9. *The graph once the edges of the augmenting chain sBDt have been calculated*

Step 2: Marking-calculation no. 4

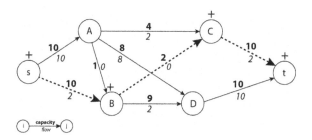

Figure 6.10. *The graph after marking no. 4 (the dotted lines are the augmenting chain)*

The marked chain is *sBCt* and has a total flow value:

$$N^+ = \min\{10 - 2 \; ; \; 2 - 0 \; ; \; 10 - 2\} = 2 \text{ and } N^- = \varnothing \Rightarrow N = 2$$

$$F = 12 + 2 = 14$$

Step 3: Updating the flow no. 4

The new flows of the edges of the augmenting chain are:

$$f(s, B) = 2 + 2 = 4 ; f(B, C) = 0 + 2 = 2 ; f(C, t) = 2 + 2 = 4$$

i.e. the graph in Figure 6.11.

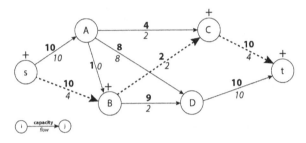

Figure 6.11. *The graph once the edges of the augmenting chain sBCt have been calculated*

Step 2: Marking-calculation no. 5

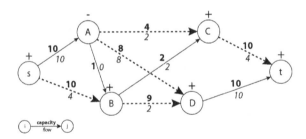

Figure 6.12. *The graph after marking no. 5 (the dotted lines are the augmenting chain)*

The marked chain is *sBDACt* and has a total flow value:

$$N^+ = \min\{10 - 4 ; 9 - 2 ; 4 - 2 ; 10 - 4\} = 2 \text{ and } N^- = \min\{ 8 \} = 8$$

$$\Rightarrow N = \min\{2 ; 8\} = 2$$

$$F = 14 + 2 = 16$$

Step 3: Updating flow no. 5

The new flows of the edges of the augmenting chain are:

f(s, B) = 4 + 2 = 6 ; f(B, D) = 2 + 2 = 4 ; f(D, A) 8 - 2 = 6 ; f(A, C) 2 + 2 = 4 ; f(C, t) 4 + 2 = 6

i.e. the graph in Figure 6.13.

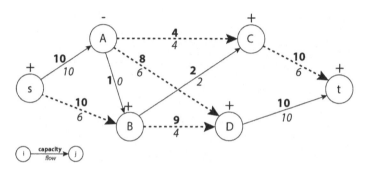

Figure 6.13. *The graph once the edges of the augmenting chain sBDACt have been calculated*

From the last graph, it is no longer possible to make any markings. Indeed, there is no longer a path from vertex *s* to vertex *t* without passing through a saturated edge. The algorithm ends and the maximum flow is therefore *F = 16.*

6.3. Minimum cut theorem

A **cut** is a subset of vertices in a network which contains at least the source *s* without the sink *p*.

Regardless of the network and the cut, its value is always higher than or equal to the value of the flow circulating in the network.

The **minimum cut** guarantees the maximum flow and confirms the calculation established by the Ford–Fulkerson algorithm.

The value of a cut is the sum of the flows of the edges exiting the cut.

6.3.1. *Example of cuts*

Let us return to our previous example and draw several cuts numbered from 1 to 4 as shown in Figure 6.14.

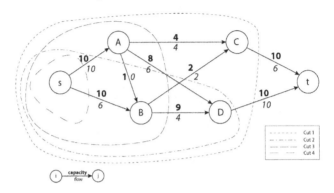

Figure 6.14. *4 cuts defined in the previous example*

The values of each of its cuts are as follows:

1) cut 1, outgoing edges (C, t) and (D, t), i.e. $6 + 10 = 16$;

2) cut 2, outgoing edges (s, A), (B, C), (D, t), i.e. $10 + 2 + 10 = 16$;

3) cut 3, outgoing edges (A, C), (A, D), (B, C) and (B, D), i.e. $4 + 6 + 2 + 4 = 16$;

4) cut 4, outgoing edges *(s, A)* and *(s, B)*, cut value : *10 + 6 = 16.*

Though they are not all represented here (so as not to complicate the representation), the minimum cut value will remain 16, which corresponds to the value of the maximum flow we have just calculated.

6.4. Dinic[3] algorithm

Whereas the Ford–Fulkerson algorithm realizes a depth-first search of the augmenting chain, giving it a linear complexity of $O(F.m)$ where F is the

---

3 Yefim Dinitz is an Israeli mathematician. His algorithm was published in 1970 from Russia, where unfortunately an error in the translation of his article led him to be known to the general public as Dinic.

maximum flow and *m* is the number of edges, Dinic's algorithm realizes a breadth-first search and its complexity is polynomial under the form of $O(m^2n)$ with *n* vertices and *m* edges.

For this algorithm we will use three types of graph: the graph *G*, the **residual graph** G_f and the **level graph** G_L .

G is the graph of origin.

The residual graph G_f is a graph where the edges *(i, j)* carry their **residual capacity** to the ongoing step, i.e. the difference of the capacity minus the flow, $c(i, j) - f(i, j)$.

The level graph G_L is a graph that positions the vertices following different levels according to their distances *d* in relation to the source *s*. It is a subgraph that only contains the edges linking a vertex to a vertex of immediately higher distance.

6.4.1. *Presentation of the algorithm*

This algorithm uses the concepts of the **augmenting path** and the **blocking flow**. An augmenting path goes from the source *s* to the sink *t* in the residual graph G_f. A blocking flow is a flow, found on an augmenting path, that contains an edge whose capacity *c(i, j)* is equal to the flow *f(i, j)*.

We consider the following variables:

s : source of the graph;

t : sink of the graph;

(i, j) : an edge;

f : flow;

F : maximum flow.

Dinic's algorithm can therefore be divided as follows:

– Initialization
attribute a flow *f* equal to *0* to all the *(i, j)*
attribute *0* to *F*

– Building the graphs and calculation
build a residual graph Gf

– Loop 1
as long as a path toward t exists
    build a level graph G_L
    find an augmenting path from the source s toward the sink t on G_f
    find the blocking flow f' on the previous augmenting path

    – Loop 2
    as long as an augmenting path exists
        find the augmenting path from the blocking flow f' until t
        find the blocking flow f' on the previous augmenting path
        update G_f (residual capacity and flow)
        calculate $F=F+f'$
    finish as long as
    update G_L
finish as long as

6.4.2. *Application on an example*

Let us return to the graph from the example used with the Ford–Fulkerson algorithm.

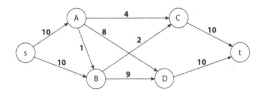

Figure 6.15. *A network for our algorithm*

Initialization:

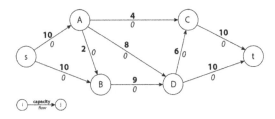

Figure 6.16. *Initialization step*

Maximum flow: *F= 0.*

Building the graphs and calculation, step 1.1.

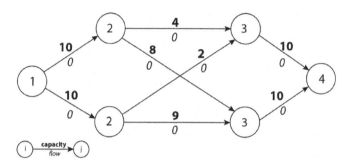

Figure 6.17. *The graph G$_L$ with its four levels*

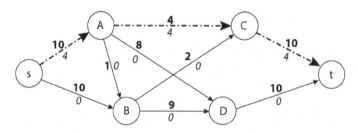

Augmenting path : {s, A, C, t}
min{f(s, A) ; f(A, C) ; f(C, t)} = {10-0 ; 4-0 ; 10-0} = 4

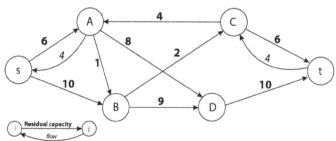

Blocking flow on (A, C)

Figure 6.18. *Graphs G and G$_f$ at the 1st passage in loops 1 and 2*

$f' = 4$ and maximum flow: $F = 0 + 4$

Building the graphs and calculation, step 1.2.

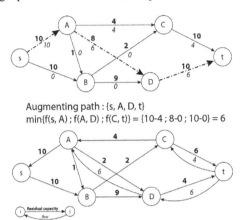

Augmenting path : {s, A, D, t}
min{f(s, A) ; f(A, D) ; f(C, t)} = {10-4 ; 8-0 ; 10-0} = 6

Blocking flow on (s, A)

Figure 6.19. *Graphs G and G_f at the first passage in loop 1 and 2nd passage in loop 2*

$f' = 6$ and maximum flow: $F = 4 + 6 = 10$.

Building the graphs and calculation, step 1.3.

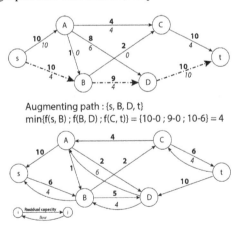

Augmenting path : {s, B, D, t}
min{f(s, B) ; f(B, D) ; f(C, t)} = {10-0 ; 9-0 ; 10-6} = 4

Blocking flow on (D, t)

Figure 6.20. *Graphs G and G_f at the first passage in loop 1 and at the 3rd passage in loop 2*

$f' = 4$ and maximum flow: $F = 10 + 4 = 14$.

Building a level graph G_L (updating).

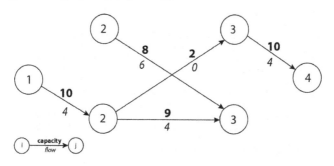

Figure 6.21. *The level graph G_L at the 2nd passage in loop 1*

Building the graphs and calculation, step 2.1.

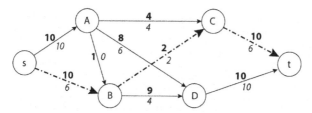

Augmenting path : {s, B, C, t}
min{f(s, B) ; f(B, C) ; f(C, t)} = {10-4 ; 2-0 ; 10-4} = 2

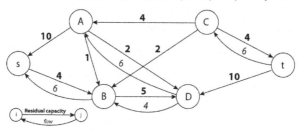

Blocking flow on (B, C)

Figure 6.22. *Graphs G and G_f at the 2nd passage in loop 1 and at the 1st passage in loop 2*

$f' = 2$ and maximum flow: $F = 14 + 2 = 16$.

There is no longer the possibility of reaching the sink t from the source s. The level graph is the same as that presented in Figure 6.23.

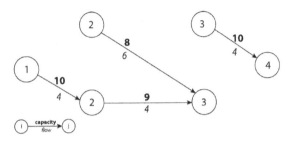

Figure 6.23. *The level graph G_L. It can be seen that the sink can no longer be reached*

The value of the maximum flow is therefore $F = 16$.

6.5. Exercises

6.5.1. *Exercise 1: drinking water supply*

The municipality of Monchâteau receives its drinking water from the network presented in Figure 6.24. The water is supplied by 3 pumping stations P1, P2 and P3 that can supply 10,000, 8000 and 6000 m³/h, and a series of lift stations R1, R2, R3, R4 and R5. Each of the water transmission pipes has the capacity indicated on the edges of the graph.

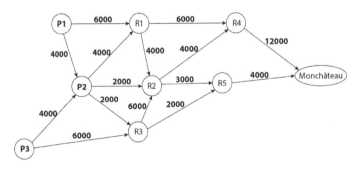

Figure 6.24. *The water distribution and supply network for the municipality of Monchâteau*

Questions:

1) What is the maximum volume of water the municipality can receive?

2) If the pumping stations provided 12,000 m³/h for P1, 10,000 m³/h for P2 and 8000 m³/h for P3, would the volume of water received by the municipality be higher than that previously obtained?

6.5.2. *Exercise 2: maximum flow according to Dinic*

Consider the network in Figure 6.25.

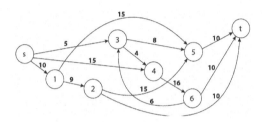

Figure 6.25. *The network for exercise 2*

Question:

Determine the maximum flow that can be transported using Dinic's algorithm.

6.5.3. *Solution to exercise 1*

Question 1:

This problem can be solved by considering it as a problem of maximum flow in a network.

To solve it, we apply the Ford–Fulkerson algorithm.

On the graph, we add a source vertex s and the municipality of Monchâteau becomes the sink t.

We then complete the graph with 3 edges that carry as a capacity the volumes supplied by the 3 pumping stations P1, P2 and P3.

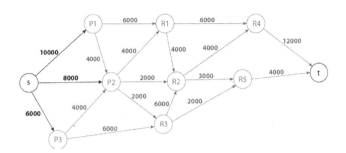

Figure 6.26. *The completed network*

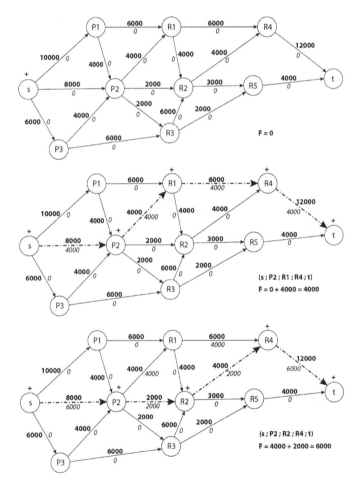

Figure 6.27. *The first 3 steps of the solution*

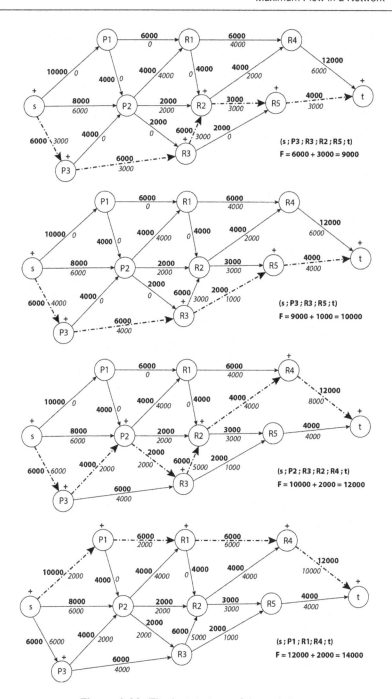

Figure 6.28. *The last 4 steps of the solution*

The maximum flow is $F = 14,000$. The municipality of Monchâteau can therefore receive $14,000\ m^3/h$.

Question 2:

The change in the volumes supplied by the pumping stations does not modify the quantity of water received (14,000 m^3/h) by the municipality of Monchâteau. This can be seen in Figure 6.29 which represents the final step of the algorithm once the paths have been traversed:

1) (s ; P1 ; R1 ; R4 ; t) ;

2) (s ; P2 ; R2 ; R4 ; t) ;

3) (s ; P3 ; R3 ; R2 ; R5 ; t) ;

4) (s ; P3 ; R3 ; R5 ; t) ;

5) (s ; P2 ; R3 ; R2 ; R4 ; t).

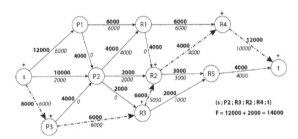

Figure 6.29. *The final step of the Ford-Fulkerson algorithm for question 2*

6.5.4. Solution to exercise 2

We start by reordering the graph and creating the associated level graph G_L.

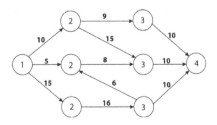

Figure 6.30. *The level graph G_L of exercise 2 once the vertices have been reordered*

We then apply the algorithm through graphs G and G_f. Each time an augmenting chain can no longer be found on a blocking flow, we update graph G_L.

Figure 6.31 shows the path, the value of flow f' and the maximum flow F.

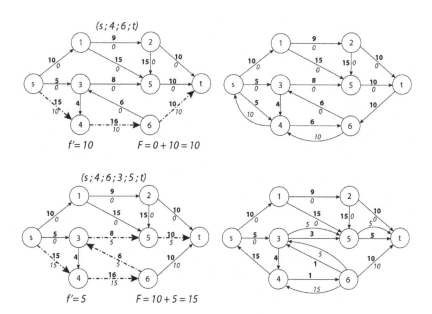

Figure 6.31. *Graphs G and the associated residual graphs G_f*

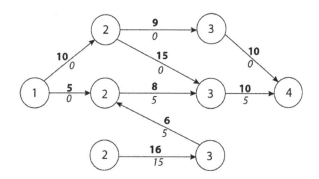

Figure 6.32. *Graph G_L*

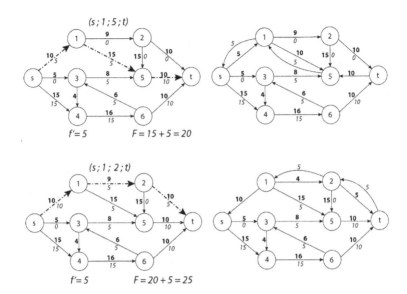

Figure 6.33. *Graphs G and the associated residual graphs G_f*

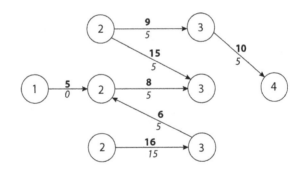

Figure 6.34. *Graph G_L*

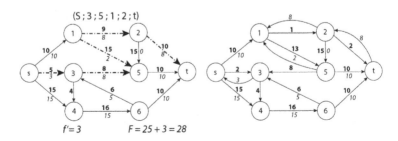

Figure 6.35. *Graphs G and the associated residual graphs G_f*

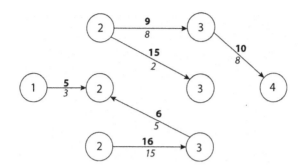

Figure 6.36. *The final graph G_L, passage is no longer possible between level 1 and level 4*

The maximum possible flow on this network is therefore: *F = 28.*

7

Trees, Tours and Transport

7.1. Fundamental concepts

When locations have known coordinates and are interconnected with paths enabling a link between one point and another potentially via intermediate points, we have a graph that is simple, connected, undirected and valued.

In this graph, we will try to define a minimum **spanning tree.**

We often encounter this type of problem in daily life, for instance, when installing a network of power lines, constructing road or rail infrastructure or installing different and varied pipe lines, among other things.

The solution lies in extracting a subgraph whose links are minimal and that ensures access to each vertex from a graph. This is an **MST** or minimum spanning tree.

Our network will consist of a graph that is simple, undirected, connected and without cycle.

Numerous algorithms exist presenting solutions to the problem, but the most commonly known are those by Kruskal[1], Prim[2] and Sollin[3].

---

1 William Henry Kruskal, 1919–2005, was an American mathematician and statistician.
2 Robert Clay Prim, born in 1921, is an American mathematician and computer scientist.
3 Following Gustave Choquet, in 1938, George Sollin rediscovered and presented, from 1961, the algorithm created by telecommunications engineer Otakar Boruvka (1899–1995) in 1926, which had never been published.

In addition to searching for an MSP, there is also the so-called traveling-salesman problem (TSP).

The TSP has fascinated countless researchers. In the early 19th Century, Hamilton[4] and Kirkman[5] were captivated by it, but it would not be until 1930 that Karl Menger, Whithney and Flood studied the problem in depth. Numerous other algorithms have since been found.

The problem can be summarized as follows:

A traveling salesman has to go to n cities once. He starts his tour in any city and finishes it by returning to the city he started from. The distances between all the cities are known. Which path should he take to minimize the distance traveled?

It is worth noting that the distance variable can be replaced by other variables, such as journey time, cost, etc.

The TSP is a representative of a category of problems called NP-complete[6]. To this day there is no known algorithm to polynomial complexity. Its complexity is $O(n!)$ with n being the number of cities. Using dynamic programming we can hope to reach a complexity of $O(n^2 2^n)^2$.

There are two types of algorithms used to solve this problem: deterministic algorithms that find an optimal solution, and approximation algorithms that provide an approximate (almost optimal) solution. The former are very complex (linked to the concept of factorials) and require high calculation power but can find the best solution. The latter generate solutions approximate to optimality in a reasonable time and with reasonable calculation power.

Deterministic algorithms can be based on the following methods:

1) brute force;

2) "branch and bound" technique;

3) dynamic programming.

---

4 Sir William Rowan Hamilton, 1805–1865, Irish mathematician, physicist and astronomer.
5 Thomas Penyngton Kirkman, 1806–1895, English mathematician.
6 Non-deterministic polynomial time (NP), a class of certain decision-making problems.

Approximation solutions use the:

1) greedy algorithm;

2) tabu algorithm;

3) simulated annealing;

4) genetic algorithms;

5) Lin-Kerningham algorithm;

6) and so on.

This chapter will present Little's algorithm which uses "branch and bound" on a simple example.

7.2. Kruskal's algorithm

This widely used algorithm is very simple to implement. It is based on the idea of the *gradient*, which consists of considering the edges of the graph in order of increasing values. Its complexity is $O(n^2 m \log m)$ for n nodes and m edges.

The following elements are considered:

$G(s, a, v)$: a graph;

v: value carried by an edge belonging to G;

A: a set of edges (lines);

P: minimum weight.

Kruskal's algorithm consists of the following steps:

– initialization;

– categorize the edges a in order of increasing values v (cost, capacity, weight, etc.);

– attribute $\emptyset$ to A.

– Solution

as long as there are edges *a* do
    if the first edge *a* does not form a cycle with the edges of *A* then
    add to *A*
end as long as

attribute to *P* the sum of the values *v* of all the edges of *A*.

7.2.1. *Application on an example*

Let us consider the graph in Figure 7.1.

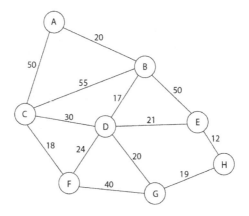

Figure 7.1. *The graph for our example*

Initialization:

    EH(12), BD(17), CF(18), GH(19), AB(20), DG(20), DE(21), DF(24), CD(30), FG(40), AC(50), BE(50), BC(55)

A = Ø

Solution:

Step 1: *A = {EH}*

Step 2: *A = {EH, BD}*

Step 3: *A = {EH, BD, CF}*

Step 4: $A = \{EH, BD, CF, GH\}$

Step 5: $A = \{EH, BD, CF, GH, AB\}$

Step 6: $A = \{EH, BD, CF, GH, AB, DG\}$

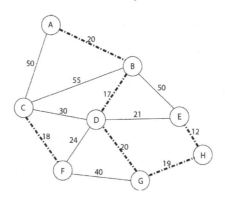

Figure 7.2. *The graph at step 6. The dotted edges have already been chosen*

Step 7: during this step, we can see that the chosen edge is *DE* which would form a cycle with the edges *EH*, *GH* and *DG*. Edge *DE* is therefore excluded and will not belong to *A*.

A = {EH, BD, CF, GH, AB, DG}

Step 8: A = {EH, BD, CF, GH, AB, DG, DF}

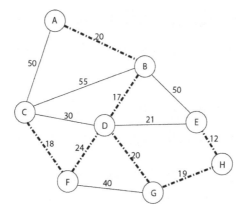

Figure 7.3. *The graph at step 8*

Step 9: during this step, we can see that the chosen edge is *CD* which would form a cycle with the edges *EH* and *DF*. *CD* is therefore excluded and will not belong to *A*.

A = {EH, BD, CF, GH, AB, DG, DF}

Step 10: the same thing happens at this step with the edge FG.

A = {EH, BD, CF, GH, AB, DG, DF}

Step 11: the same thing happens at this step with the edge AC.

A = {EH, BD, CF, GH, AB, DG, DF}

Step 12: the same thing happens at this step with the edge BE.

A = {EH, BD, CF, GH, AB, DG, DF}

Step 13: the same thing happens at this step with the edge BC.

A = {EH, BD, CF, GH, AB, DG, DF}

The graph in Figure 7.3 shows the MST:

$$P = v_{EH} + v_{BD} + v_{CF} + v_{GH} + v_{AB} + v_{DG} + v_{DF}$$

$$P = 12 + 17 + 18 + 19 + 20 + 20 + 24 = 130$$

COMMENT 7.1.– At step 8, it can be seen that all the vertices have been reached and the minimum tree has been obtained and the algorithm can therefore end here. The number of edges in a minimum spanning tree is always equal to n-1 vertices.

7.3. Prim's algorithm

Prim's algorithm is similar to Dijkstra's[7] which is used to calculate the shortest paths. We select the edge with the minimum weight from

---

7 See Chapter 3, section 3.2.

the vertices that are already marked in the tree and the adjacent vertices that are not yet part of the tree. At the start, the tree consists of a vertex chosen at random. Its complexity is $O(n^2)$.

The following elements are considered:

$G(s, a, p)$: a graph;

i, j : ends of the edge a;

p: weight carried by an edge (i, j) ;

n: total number of vertices s;

N: total number of marked vertices;

R: a set of edges (lines);

P: total minimum weight.

Prim's algorithm consists of the following steps:

– Initialization

attribute 0 to p
attribute 0 to P
attribute $\emptyset$ to R
attribute 1 to N
choose any vertex s at random from the graph and mark it

– Solution

as long as N is lower to n do:
– look for the edge with the minimum weight (i, j) adjacent to this vertex (or the previously defined tree)
– mark this edge $a(i, j)$ and its end vertex j
– add to R
– attribute $N + 1$ to N
– attribute $P + p$ to P
end as long as
return R and P.

7.3.1. *Application on an example*

Let us return to the previous example shown in Figure 7.1.

– Initialization:

$p = 0$; $P = 0$; $R = \emptyset$; $N = 1$

chosen and marked vertex: A

– Solution:

Step 1: $N = 1$ and $n = 8$

mark edge AB

mark vertex B

$\quad R = \{AB\}$

$\quad N = 2$

$\quad P = 0 + 20 = 20$

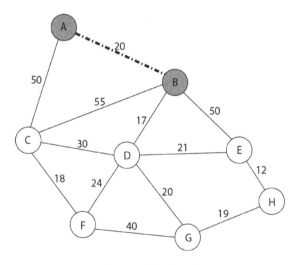

Figure 7.4. *Step 1*

Step 2: $N = 2$ and $n = 8$

mark edge BD

mark vertex D

   $R = \{AB, BD\}$

   $N = 3$

   $P = 20 + 17 = 37$

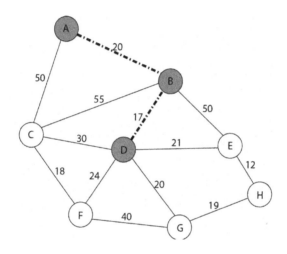

Figure 7.5. *Step 2*

Step 3: $N = 3$ and $n = 8$

mark edge DG

mark vertex G

   $R = \{AB, BD, DG\}$

   $N = 4$

   $P = 37 + 20 = 57$

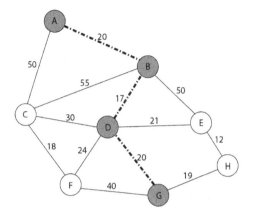

Figure 7.6. *Step 3*

Step 4: *N = 4* and *n = 8*

mark edge *GH*

mark vertex *H*

$R = \{AB, BD, DG, GH\}$

$N = 5$

$P = 57 + 19 = 76$

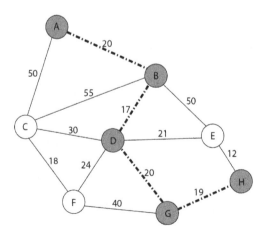

Figure 7.7. *Step 4*

Step 5: $N = 5$ and $n = 8$

mark edge *HE*

mark vertex E

$R = \{AB, BD, DG, GH, HE\}$

$N = 6$

$P = 76 + 12 = 88$

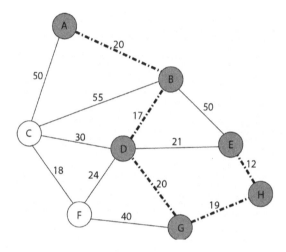

Figure 7.8. *Step 5*

Step 6: $N = 6$ and $n = 8$

mark edge *DF*

mark vertex *F*

$R = \{AB, BD, DG, GH, HE, DF\}$

$N = 7$

$P = 88 + 24 = 112$

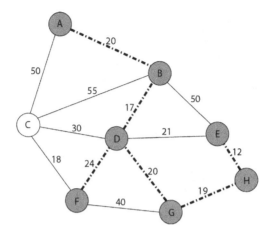

Figure 7.9. *Step 6*

Step 7: *N = 7* and *n = 8*

mark edge *FC*

mark vertex C

$R = \{AB, BD, DG, GH, HE, DF, FC\}$

$N = 7$

$P = 112 + 18 = 130$

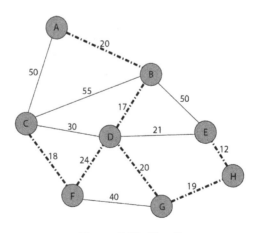

Figure 7.10. *Step 7*

$N = 8$ and $n = 8$

$R = \{AB, BD, DG, GH, HE, DF, FC\}$

$P = 130$

7.4. Sollin's algorithm

A hybrid of Kruskal and Prim's algorithms, Sollin's algorithm uses optimality conditions linked to cuts. It defines a set of spanning trees to which the minimum weight edges are added. Its complexity is $O(m \log_2 n)$ for n nodes and m edges.

The algorithm is based around two main steps: first initialization and then the construction of subtrees (connected components of the minimum weight tree under construction) by marking vertices and edges, interconnecting these subtrees and then marking them to generate a MST.

The following elements are considered:

$G(s, a, p)$: a graph;

p: weight carried by an edge a;

n: total number of vertices s;

N: total number of marked vertices;

P: minimum total weight;

R: a set of edges (lines).

The various steps of the algorithm are as follows:

– Initialization

attribute to n the number of vertices in the graph
attribute 0 to p
attribute 0 to P
attribute $\varnothing$ to R

choose a vertex *s* at random

mark *s*

– Solution

as long as *n* is higher than *1* do

     as long as there is an unmarked vertex do

          look for the edge with the minimum weight *a* adjacent to the previously marked *s*

          add to *R*

          mark the edge *a*

          mark the end vertex *s* of this edge *a* (if it is not already)

          attribute *P* + *p* to *P*

          attribute *n* – *1* to *n*

          select a new unmarked vertex *s* at random

          mark *s*

     end as long as

     select an unmarked sub-graph *g* at random

     mark *g*

     look for the edge with the minimum weight *a* adjacent to previously marked *g*

     add to *R*

     mark the edge *a*

     mark the subgraph *g* end of this edge *a* (if it is not already)

     attribute *P* + *p* to *P*

     attribute *n* – *1* to *n*

end as long as

return *R* and *P*.

7.4.1. *Application on an example*

Let us return again to the example shown in Figure 7.1.

– Initialization:

   n = 8 ; p = 0 ; P = 0 ; R = Ø

chosen vertex: *A*

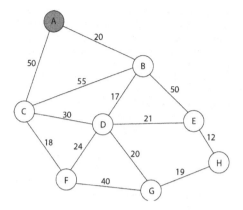

Figure 7.11. *Initialization*

Solution:

Step 1:

edge with minimum weight: *AB*

$R = \{AB\}$

$P = 0 + 20 = 20$

$n = 8 - 1 = 7$

chosen vertex: *C*

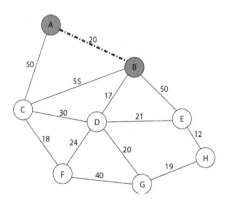

Figure 7.12. *Step 1*

Step 2:

edge with minimum weight: *CF*

$R = \{AB, CF\}$

$P = 20 + 18 = 38$

$n = 7 - 1 = 6$

chosen vertex: *D*

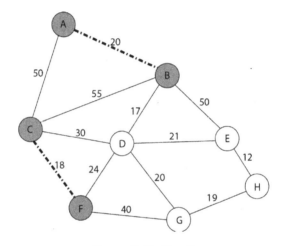

Figure 7.13. *Step 2*

Step 3:

edge with minimum weight: *DB*

$R = \{AB, CF, DB\}$

$P = 38 + 17 = 55$

$n = 6 - 1 = 5$

chosen vertex: *E*

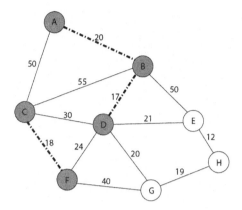

Figure 7.14. *Step 3*

Step 4:

edge with minimum weight: *EH*

    R = {AB, CF, DB, EH}

    P = 55 + 12 = 67

    n = 5 – 1 = 4

chosen vertex: *G*

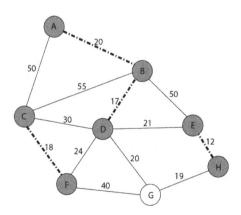

Figure 7.15. *Step 4*

Step 5:

edge with minimum weight: *EH*

   $R = \{AB, CF, DB, EH\}$

   $P = 55 + 12 = 67$

   $n = 5 - 1 = 4$

chosen vertex: *G*

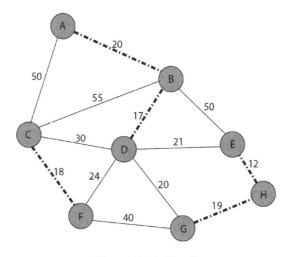

Figure 7.16. *Step 5*

Step 6:

edge with minimum weight: *GH*

   $R = \{AB, CF, DB, EH, GH\}$

   $P = 67 + 19 = 86$

   $n = 4 - 1 = 3$

chosen subgraph: *ABD*

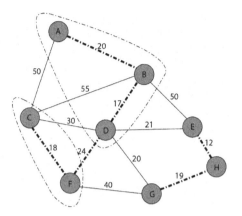

Figure 7.17. *Step 6*

Step 7:

edge with minimum weight: *FD*

> *R = {AB, CF, DB, EH, GH, FD}*
>
> *P = 86 + 24 = 110*
>
> *n = 3 – 1 = 2*

chosen subgraph: *EHG*

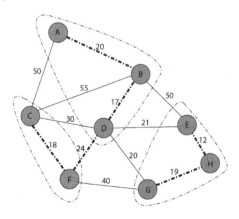

Figure 7.18. *Step 7*

Step 8:

edge with minimum weight: *DG*

$R = \{AB, CF, DB, EH, GH, FD, DG\}$

$P = 110 + 20 = 130$

$n = 2 - 1 = 1$

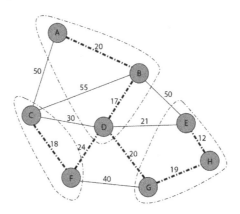

Figure 7.19. *Step 8*

$n = 1$

$R = \{AB, CF, DB, EH, GH, FD, DG\}$

$P = 110 + 20 = 130$

7.5. Little's algorithm for solving the TSP

The algorithm was published by John D.C. Little[8] in March 1963. As has already been mentioned, this algorithm is part of the "branch and bound" family and as such generates an optimal solution to the problem asked. It must however be reserved for problems concerning a small number of cities to obtain a solution that uses a reasonable amount of resources.

---

8 John Dutton Conant Little, born in 1928, is an American academic, professor at MIT and specialist in operational research.

This algorithm is based around two main steps: initialization and "branch and bound". We start with a matrix that will be reduced while constructing a search tree from a father node that will define the optimal route. By the end of the algorithm, a route solution is found which may be improved by backtracking the search tree.

The steps of the algorithm are as follows:

– Initialization

Reduction of the distance (or cost) matrix
    reduce the line: remove the smallest element in each line
    reduce the column: remove the smallest element in each column

    determine of the father node and its cost (sum of lines + columns reductions)

– Branch and bound

as long as it is possible for the new matrix do
-   calculate the regrets
-   select the edge with the maximum regret
-   make a branch to create the arborescence
-   right branch (included) if this edge is traversed
-   left branch (excluded) if this edge is not traversed
-   determine the cost of the son node of the left branch (cost of the father + regret)
-   create the new matrix
-   remove the edge creating a loop, where it exists (sub-tour).
-   reduce the line: remove the smallest element in each line
-   reduce the column: remove the smallest element in each column
-   calculate the sum of the lines + columns reductions
-   determine the cost of the son node of the right branch (previous total + cost of the father)

end as long as

-   add the missing route(s) to conclude the tour;
-   return to the primary solution (total cost of the route = cost of the final node of the right branch);
- backtrack the search tree and check the cost of the nodes of the excluded branches.

7.5.1. *Application on an example*

Figure 7.20 shows a graph representing the distances between a number of French cities.

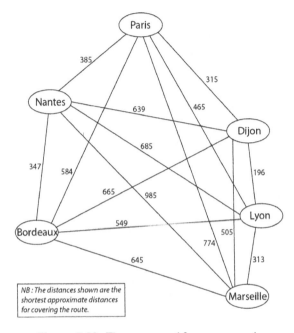

Figure 7.20. *The map used for our example*

We formalize this graph as a symmetric matrix (in our case, we consider that the distances from one city to another are identical regardless of the direction of the journey connecting them).

Cities	Paris (P)	Dijon (D)	Lyon (L)	Marseille (M)	Bordeaux (B)	Nantes (N)
Paris (P)		315	465	774	584	385
Dijon (D)	315		196	503	665	639
Lyon (L)	465	196		313	549	685
Marseille (M)	774	503	313		645	985
Bordeaux (B)	584	665	549	645		347
Nantes (N)	385	639	685	985	347	

Table 7.1. *The symmetric matrix*

We apply the algorithm:

Initialization: Reduce the line and the column.

Cities	Bordeaux	Lyon	Nantes	Paris	Marseille	Dijon	Min
Bordeaux		549	347	584	645	665	347
Lyon	549		685	465	313	196	196
Nantes	347	685		385	985	639	347
Paris	584	465	385		774	315	315
Marseille	645	313	985	774		503	313
Dijon	665	196	639	315	503		196

1714

Table 7.2. *Search for the minimums in each line*

For each cell in a line, we remove the corresponding minimum value found.

Cities	Bordeaux	Lyon	Nantes	Paris	Marseille	Dijon
Bordeaux		202	0	237	298	318
Lyon	353		489	269	117	0
Nantes	0	338		38	638	292
Paris	269	150	70		459	0
Marseille	332	0	672	461		190
Dijon	469	0	443	119	307	
Min	0	0	0	38	117	0

155

Table 7.3. *Reduction of the matrix according to the minimums in each line and search for the minimums in each column*

For each cell in a column, we remove the corresponding minimum value found.

Cities	Bordeaux	Lyon	Nantes	Paris	Marseille	Dijon
Bordeaux		202	0	199	181	318
Lyon	353		489	231	0	0
Nantes	0	338		0	521	292
Paris	269	150	70		342	0
Marseille	332	0	672	423		190
Dijon	469	0	443	81	190	

Table 7.4. *Reduction of the matrix according to the minimums in each column*

Determining the cost carried by father node 1:

The sum of the minimum lines + columns gives the value of the father node by default, i.e. *1714 + 155 = 1869*.

Search tree

1 (1869)

Figure 7.21. *The father vertex of our search tree*

Calculate the **regrets**. The regret of an edge, in this case the distance between two cities, is the sum of the smallest value in the line and the smallest value in the column.

The calculation of the regrets is sometimes called the **calculation of avoidance**.

For example:

1) For the edge *BN* (Bordeaux – Nantes), we have a minimum value of *181* for the line and a minimum value of *70* for the column, i.e. *181 + 70 = 251*.

2) For the edge *PD* (Paris – Dijon), we have a minimum value of *70* for the line and a minimum value of *0* for the column, i.e. *70 + 0 = 70*.

3) For the edge *ML* (Marseille – Lyon), we have a minimum value of *190* for the line and a minimum value of *0* for the column, i.e. *190 + 0 = 190*.

Continuing in the same way for each of the cells containing a value of null, we generate the matrix in Figure 7.25.

Cities	Bordeaux	Lyon	Nantes	Paris	Marseille	Dijon
Bordeaux		202	251	199	181	318
Lyon	353		489	231	181	0
Nantes	**269**	338		81	521	292
Paris	269	150	70		342	70
Marseille	332	190	672	423		190
Dijon	469	81	443	81	190	

Table 7.5. *Calculation of the regrets*

The search for the route with the maximum regret is in our case: *NB* = *269*.

COMMENT 6.2.– If more than one regret has the same value, one is chosen at random.

We create two branches from father node 1 which correspond to the exclusion or inclusion of the route in our future tour.

We determine the cost of son node 2 in the excluded left branch. It is equal to the sum of the cost of the father node and the regret, i.e. *1869 + 269* = *2138*.

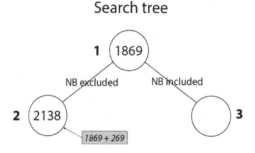

Search tree

Figure 7.22. Creation of the branches and nodes 2 and 3 from father node 1

We create a new matrix by removing the line and the column corresponding to the route *NB* (previous maximum regret).

Cities	Lyon	Nantes	Paris	Marseille	Dijon
Bordeaux	202	0	199	181	318
Lyon		489	231	0	0
Paris	150	70		342	0
Marseille	0	672	423		190
Dijon	0	443	81	190	

Table 7.6. The new matrix

To avoid a loop (subtour: *NB-BN*), we remove the route *BN*.

Cities	Lyon	Nantes	Paris	Marseille	Dijon
Bordeaux	202		199	181	318
Lyon		489	231	0	0
Paris	150	70		342	0
Marseille	0	672	423		190
Dijon	0	443	81	190	

Table 7.7. *Removing the loop BN (subtour)*

We reduce the new matrix.

Cities	Lyon	Nantes	Paris	Marseille	Dijon	**Min**
Bordeaux	202		199	181	318	181
Lyon		489	231	0	0	0
Paris	150	70		342	0	0
Marseille	0	672	423		190	0
Dijon	0	443	81	190		0

181

Table 7.8. *Search for the minimums in each line*

Cities	Lyon	Nantes	Paris	Marseille	Dijon
Bordeaux	21		18	0	137
Lyon		489	231	0	0
Paris	150	70		342	0
Marseille	0	672	423		190
Dijon	0	443	81	190	
Min	0	70	18	0	0

88

Table 7.9. *Reduction of the matrix according to minimums in each line and search for the minimums in each column*

Calculate the cost of son node 3 in the included right branch.

Sum of the lines + column reductions, i.e. *181 + 88 = 269.*

Cost of the father: *1869*.

Cost of the son: *1869 + 269 = 2138*.

Search tree

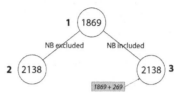

Figure 7.23. *Son node 3 in the included branch and its cost*

The "branch and bound" step is then repeated.

Cities	Lyon	Nantes	Paris	Marseille	Dijon
Bordeaux	21		63	0	137
Lyon		419	213	0	0
Paris	150	**373**		342	0
Marseille	190	602	405		190
Dijon	63	373	63	111	

Table 7.10. *Calculation of the regrets*

Maximum regret *PN = 373*.

Cost of the new father node 3: *2138*.

Cost of the son node 4 in the excluded branch: *2138 + 373 = 2511*.

Cities	Lyon	Paris	Marseille	Dijon
Bordeaux	21		0	137
Lyon		150	0	0
Marseille	0	342		190
Dijon	0	0	111	

Table 7.11. *New matrix with the removal of the subtour BP and reduction*

Sum of the lines + column reductions: $0 + 63 = 63$.

Cost of the son node 5 in the included branch: $2138 + 63 = 2201$.

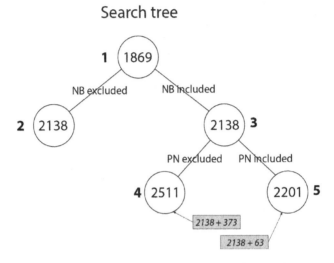

Figure 7.24. *Son nodes 4 and 5 and their costs*

Remove edge *BP* to avoid a loop (subtour: *PN-NB-BP*).

Cities	Lyon	Paris	Marseille	Dijon
Bordeaux	21		21	137
Lyon		150	0	137
Marseille	**190**	342		190
Dijon	63	150	111	

Table 7.12. *Calculation of the regrets*

Maximum regret ML = 190.

Cost of the new father node 5: 2201.

Cost of the son node 6 in the excluded branch: 2201 + 190 = 2391.

Cities	Paris	Marseille	Dijon
Bordeaux		0	137
Lyon	150		0
Dijon	0	111	

Table 7.13. *New matrix with the removal of subtour LM and reduction*

Sum of the lines + columns reductions: *0 + 0 = 0*.

Cost of the son node 7 in the included branch: *2201 + 0 = 2201*.

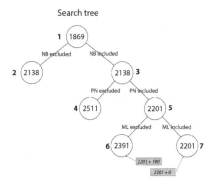

Figure 7.25. *Son nodes 6 and 7 and their costs*

Remove edge *LM* to avoid a loop (subtour: *ML-LM*).

Cities	Paris	Marseille	Dijon
Bordeaux		248	137
Lyon	150		**287**
Dijon	261	111	

Table 7.14. *Calculation of the regrets*

Maximum regret LD = 287.

Cost of the new father node 7: 2201.

Cost of the son node 8 in the excluded branch: $2201 + 287 = 2488$.

Cities	Paris	Dijon
Bordeaux		0
Dijon	0	111

Table 7.15. *New matrix and reduction*

Sum of the lines + column reductions: *0 + 0 = 0*.

Cost of the son node 9 in the included branch: *2201 + 0 = 2201*.

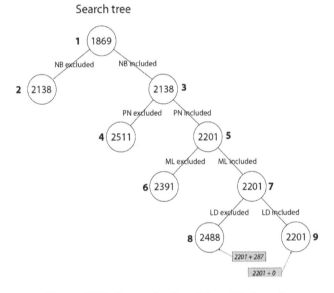

Figure 7.26. *Son nodes 8 and 9 and their costs*

Cities	Paris	Marseille
Bordeaux		111
Dijon	**111**	111

Table 7.16. *Calculation of the regrets*

Maximum regret *DP* (the same values imposes a random choice, we could have also taken *BM*) = *111*.

Cost of the new father node 9: *2201*.

Cost of the son node 10 in the excluded branch: *2201 + 111 = 2312*.

Sum of the lines + column reductions: *0 + 0 = 0*.

Cost of the son node 11 in the included branch: *2201 + 0 = 2201*.

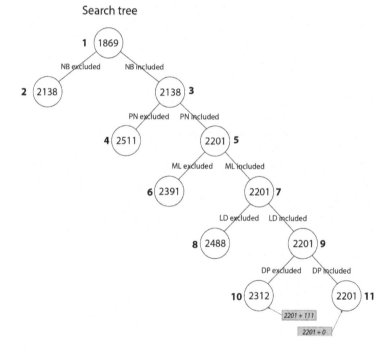

Figure 7.27. *Son nodes 10 and 11 and their costs*

Add the route *BM* to close the tour which becomes: *PN-NB-BM-ML-LD-DP*.

Primary solution: *total cost = 2201*.

All the nodes in the excluded branches are higher than *2201*, the primary solution is therefore optimal.

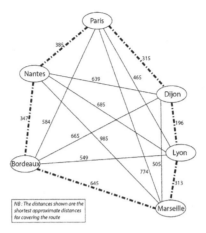

Figure 7.28. *The solution to our example, the shortest tour for visiting all the cities*

7.6. Exercises

7.6.1. *Exercise 1: Computer network*

We want to build a computer network connecting a set of industrial buildings located across a wide geographical area. The cabling will use fiber optics installed in casings buried underground. The work involved to bury and connect the cables is not insignificant and comes at a high cost. Following a study on the ground, the different possible link options have been presented in the layout diagram in Figure 7.44.

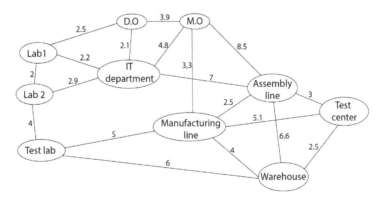

Figure 7.29. *A diagram of the potential layout for the network containing the overall cost of the work (burying, installing the fiber, connecting, materials, workforce, etc.) corresponding, for each link, to thousands of euros*

We need to design an architecture that minimizes the financial costs of the project by offering the best solution.

Question:

Determine, using an appropriate algorithm, an optimal solution and its corresponding cost.

7.6.2. Exercise 2: Delivery

Traveling in a van, a courier must make deliveries to 6 shops (S1–S6) located in the town center. The distances between the shops are laid out in Table 7.17. The distances are expressed in meters.

The routes are not symmetric. The presence of countless one-way streets in the town gives different backtrack routes. For example *B4* toward *B3* = *1000 m* whereas *B3* toward *B4* = *600 m*.

Shop	S1	S2	S3	S4	S5	S6
S1		200	1400	600	2800	400
S2	600		1200	1800	200	2400
S3	1200	2800		600	1400	600
S4	400	600	1000		1800	2200
S5	3000	1400	2200	400		800
S6	4000	1000	2600	800	3600	

Table 7.17. *Distances between the shops in meters*

Question:

Determine the tour that optimizes the length of this courier's journey.

7.6.3. Solution to exercise 1

This problem can be resolved by Kruskal's, Prim's or Sollin's algorithm.

The solution using Kruskal's algorithm is laid out below.

First we reorganize the layout diagram to make it more comprehensible and better suited for processing. Each vertex is replaced by an equivalent marker using the letters of the alphabet:

– Lab 1: A;

– Lab 2: B;

– Test lab: C;

– D.O. (Design office): D;

– Info service: E;

– Methods office (M.O.): F;

– Manufacturing line: G;

– Assembly line: H;

– Warehouse: I;

– Test center: J.

We thus obtain the graph shown in Figure 7.30.

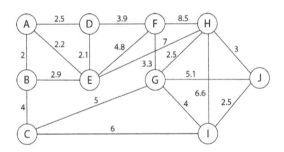

Figure 7.30. *The graph obtained after reorganization*

Initialization:

Let: AB(2) ; DE(2,1) ; AE(2,2) ; AD(2,5) ; IJ(2,5) ; GH(2,5) ; BE(2,9) ; HJ(3) ; FG(3,3) ; DF(3,9) ; BC(4) ; GI(4) ; EF(4,8) ; CG(5) ; GJ(5,1) ; CI(6) ; HI(6,6) ; EH(7) ; FH(8,5)

A = ∅

Solution:

Step 1 : *A* = *{AB}*

Step 2 : *A* = *{AB, DE}*

Step 3 : *A* = *{AB, DE, AE}*

Step 4 : Removal of *AD* (loop with *AE-DE*, therefore *A* = *{AB, DE, AE}*

Step 5 : *A* = *{AB, DE, AE ; IJ}*

Step 6 : *A* = *{AB, DE, AE, IJ, GH}*

Step 7 : Removal of *BE* (loop with *B-AE*), therefore *A* = *{AB, DE, AE ; IJ ; GH}*

Step 8 : *A* = *{AB, DE, AE ,IJ ,GH, HJ}*

Step 9 : *A* = *{AB, DE, AE ,IJ ,GH, HJ, FG}*

Step 10 : *A* = *{AB, DE, AE ,IJ ,GH, HJ, FG ,DF}*

Step 11 : *A* = *{AB, DE, AE ,IJ ,GH, HJ, FG ,DF, BC}*

Step 12 : Removal of *GI* (Loop with *GH-HJ-IJ*), therefore *A* = *{AB, DE, AE, IJ ,GH, HJ, FG ,DF, BC}*

Step 13 : Removal of *EF* (Loop with *DE-DF*), therefore *A* = *{AB, DE, AE, IJ ,GH, HJ, FG ,DF, BC}*

Step 14 : Removal of *GC* (Loop with *BC-BE-EF-FG*), therefore *A* = *{AB, DE, AE, IJ ,GH, HJ, FG ,DF, BC}*

Step 15 : Removal of *GJ* (Loop with *GH-HJ*), therefore *A* = *{AB, DE, AE, IJ ,GH, HJ, FG ,DF, BC}*

Step 16 : Removal of *CI* (Loop with *CG-GI*), therefore *A* = *{AB, DE, AE, IJ ,GH, HJ, FG ,DF, BC}*

Step 17 : Removal of *HI* (Loop with *HJ-IJ*), therefore *A* = *{AB, DE, AE, IJ ,GH, HJ, FG ,DF, BC}*

Step 18 : Removal of *EH* (Loop with *BE-BC-CG-GH*), therefore *A* = *{AB, DE, AE, IJ ,GH, HJ, FG ,DF, BC}*

Step 19 : Removal of *EH* (Loop with *EF-FH*), therefore *A* = *{AB, DE, AE, IJ ,GH, HJ, FG ,DF, BC}*

The graph shown in Figure 7.31 represents the MST.

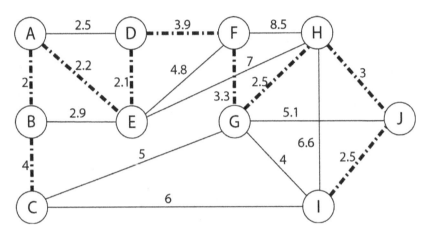

Figure 7.31. *The minimum spanning tree providing a solution to exercise 1*

$$P = v_{AB} + v_{DE} + v_{AE} + v_{IJ} + v_{GH} + v_{HJ} + v_{FG} + v_{DF} + v_{BC}$$

$$P = 2 + 2{,}1 + 2{,}2 + 2{,}5 + 2{,}5 + 3 + 3{,}3 + 3{,}9 + 4 = 25{,}50 \text{ k€}$$

COMMENT 6.3.– At step 11 we can see that the vertices have been reached, the minimum tree has already been obtained and the algorithm can finish here.

7.6.4. *Solution to exercise 2*

Here only the reduced matrices, the matrices of the calculation of the regrets and the search tree are present; this should be sufficient to correctly realize a solution.

Shop	S1	S2	S3	S4	S5	S6
S1		0	600	400	2600	200
S2	400		400	1600	0	2200
S3	600	2200		0	800	0
S4	0	200	0		1400	1800
S5	2600	1000	1200	0		400
S6	3200	200	1200	0	2800	

Table 7.18. *Reduction no. 1*

Shop	S1	S2	S3	S4	S5	S6
S1		**400**	600	400	2600	200
S2	400		400	1600	**1200**	2200
S3	600	2200		0	800	**200**
S4	**400**	200	**400**		1400	1800
S5	2600	1000	1200	**400**		400
S6	3200	200	1200	**200**	2800	

Table 7.19. *Calculation of the regrets no. 1*

Shop	S1	S2	S3	S4	S6
S1		0	600	400	200
S2	600	2200		0	0
S3	0	200	0		1800
S4	2600		1200	0	400
S6	3200	200	1200	0	

Table 7.20. *Reduction no. 2*

Shop	S1	S2	S3	S4	S6
S1		**400**	600	400	200
S2	600	2200		0	**200**
S3	**600**	200	**600**		1800
S4	2600		1200	**400**	400
S6	3200	200	1200	**200**	

Table 7.21. *Calculation of the regrets no. 2*

Shop	S2	S3	S4	S6
S2	0	0		200
S3	2200		0	0
S4		600	0	400
S6	200	600	0	

Table 7.22. *Reduction no. 3*

Shop	S2	S3	S4	S6
S2	**200**	**600**		200
S3	2200		0	**200**
S4		600	**400**	400
S6	200	600	**200**	

Table 7.23. *Calculation of the regrets no. 3*

The route S2S5 included – S4S1 included – S1S2 excluded (4400) is higher than S2S5 included – S4S1 included – S1S3 (4000) included in the search tree. We must therefore return to the matrix before line S4 and S1 have been removed.

Shop	S1	S2	S3	S4	S6
S1		0	600	400	200
S2	0	2200		0	0
S3		200	0		1800
S4	2000		1200	0	400
S6	2600	200	1200	0	

Table 7.24. *Reduction no. 4 of the matrix for the calculation of the branch S3S1 excluded and S3S1 included*

Shop	S1	S2	S3	S4	S6
S1		**400**	600	400	200
S2	**2000**	2200		0	**200**
S3		200	**800**		1800
S4	2600		1200	**400**	400
S6	3200	200	1200	**200**	

Table 7.25. *Calculation of the regrets no. 4*

Shop	S2	S4	S6
S2	2000		0
S4		0	400
S6	0	0	

Table 7.26. *Reduction no. 5*

Shop	S2	S4	S6
S2	2000		**2400**
S4		**400**	400
S6	**2000**	0	

Table 7.27. *Calculation of the regrets no. 5*

Shop	S2	S4
S2		0
S4	0	0

Table 7.28. *The final matrix, we can add S5S4 and S6S2 without increasing the cost*

Search tree

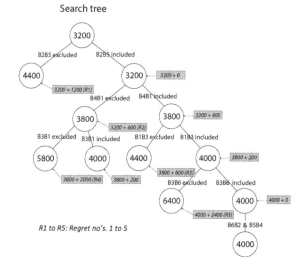

R1 to R5: Regret no's. 1 to 5

Figure 7.32. *The global search tree, the solution to our exercise*

The final solution is the tour: S2S5S4S1S3S6S2 and the total length of the journey is 4000 m.

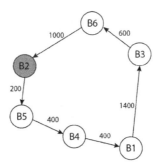

Figure 7.33. *The tour S2S5S4S1S3S6S2*

8

Linear Programming

8.1. Fundamental concepts

Linear programming determines the optimal use of a resource to maximize or minimize a cost. It is based on a mathematical technique following three methods[1]:

– a graphic solution;

– an algebraic solution;

– the use of the **simplex** algorithm.

Using a graphic solution is restrictive as it can only manage 2 or 3 variables.

Simplex exists to simplify the algebraic method for which calculations quickly become complex.

There are three steps for solving a linear programming problem:

– **identification** which decrypts and identifies the data of the problem;

– **formalization** which consists of modeling the problem in the form of linear equations or inequations to define and form the different constraints;

– solution which looks for the optimum to bring about the final solution.

---

1 There is a fourth method. It is also algebraic and is based on Cramer's solution of linear systems. However, it cannot always be used as the number of equations needing to be processed is often too high.

In this chapter, we will study the graphic method and the simplex method on two simple examples before implementing them in a number of exercises. We will then study duality, which associates with a linear programming problem, known as a **primal** problem, a second problem, known as a **dual** problem. The latter is inextricably linked to the former; the optimality of one is entirely determined by the solution of the other.

8.1.1. *Formulating a linear program*

Before we start, let us remind ourselves of the general formulation.

There are three variables c, b and A

$$c \in \mathbb{R}^n, b \in \mathbb{R}^m \ and \ A \in \mathbb{R}^{n \times m}$$

$$max \sum_{j=1}^{n} c_j x_j \ \text{with} \ max_{x \in \mathbb{R}^n} \ \ c^T x$$

under constraints

$$\sum_{j=1}^{n} a_{ij} x_j \leq b_i, i = 1, ..., m \ \text{with} \ Ax \leq b \ \text{and} \ x \geq 0$$

$$x_j \geq 0, j = 1, ..., n$$

For maximization, all the inequalities have the direction $\leq$ and the direction $\geq$ for the minimizations.

Each constraint has a constant part which can be found on the right side of the inequation.

When all these conditions are combined, it is said that the linear program is in its **standard form**.

8.2. Graphic solution method

In this method, we will first delineate a domain with the half-plane intersections which represent the inequations of the constraints. We then study the borders of this domain to determine the points providing the optimum of the function.

We will demonstrate how this technique works with an example.

A factory manufactures two products P1 and P2. Three production floors F1, F2 and F3 are used to make a product. The hourly consumption of energy needed to make each product is known and limited.

The table below brings together the data.

Hourly consumption of energy			
Floor	Product P1	Product P2	Limitation
F1	3	6	<=21 kwh
F2	3	3	<=14 kwh
F3	3	0	<=10.5 kwh

Table 8.1. *The data for the example*

For each product sold, the manufacturer makes a profit of $6.00 on the sale of product P1 and a profit of $4.00 on the sale of product P2.

We want to find out the hourly production for each product which maximizes the profit for the manufacturer while taking into account all the constraints.

8.2.1. Identification

We call q_1 and q_2 the quantities of products manufactured each hour by each of the floors and B the profit.

8.2.2. Formalization

We will first define the constraints in the form of inequations that will have an effect on q_1, q_2 and the limitation:

$$3q_1 + 6q_2 <= 24$$

$$3q_1 + 3q_2 <= 15$$

$$3q_1 <= 12$$

We can also write (as the quantity produced is always positive):

$$q_1 >= 0 \text{ and } q_2 >= 0$$

To conclude, we will return to the question asked in the problem statement: what we are looking to maximize is profit B, i.e.:

$$B = 6q_1 + 4q_2$$

We will now bring back the inequations and place them in an orthonormal frame of reference with q_1 as the abscissa and q_2 as the ordinate.

All the possible solutions for this problem will correspond to the surface formed by all the points meeting the constraints.

In Figures 8.1–8.5 we have plotted each of the 5 highest constraints.

The points to be kept are indicated by the light part of the graph.

Constraint: $3q_1 + 6q_2 <= 24$.

We transform the inequation into an equation to define the straight line bordering the half-plane, i.e. *$3q_1 + 6q_2 = 24$*.

We calculate 2 points of this straight line so as to be able to plot it:

With $q_1 = 0$, we have: $q_2 = \dfrac{-3 \times 0 + 24}{6} = \dfrac{24}{6} = 4$

With $q_1 = 2$, we have: $q_2 = \dfrac{-3 \times 2 + 24}{6} = \dfrac{18}{6} = 3$

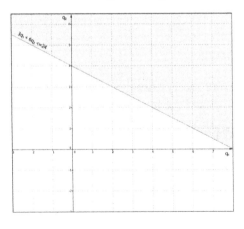

Figure 8.1. *Plot of the 1ˢᵗ half-plane defined by the constraint $3q_1 + 6q_2 <= 24$. The straight line passes through the pair of coordinates (0, 4) and (2, 3)*

COMMENT 8.1.– To determine which half-plane, below or above the straight line, should be kept to form the set of the solutions, you take a pair of coordinates from either side, for instance, $(2, 2)$ and use them in the inequation of the constraint. If verified, the pair is part of the half-plane to be kept, otherwise the opposite is true.

Let $3q_1 + 6q_2 <= 24$, with the pair $(2, 2)$, we have:

$$3\times2 + 6\times2 <= 24 \implies 6 + 12 <= 24 \implies 18 <= 24$$

The inequality is verified and therefore the pair $(2, 2)$ belongs to the half-plane of the solutions.

Constraint: $3q_1 + 3q_2 <= 15$.

We transform the inequation into an equation to define the straight line bordering the half-plane, i.e. $3q_1 + 3q_2 = 15$.

We calculate 2 points of this straight line so as to be able to plot it:

With $q_1 = 1$, we have: $q_2 = \dfrac{-3\times1+15}{3} = \dfrac{12}{3} = 4$

With $q_1 = 3$, we have: $q_2 = \dfrac{-3\times3+15}{3} = \dfrac{6}{3} = 2$

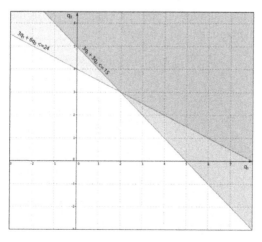

Figure 8.2. *Plot of the 2nd half-plane defined by the constraint $3q_1 + 3q_2 <= 15$. The straight line passes through the pairs of coordinates (1, 4) and (3, 2)*

Constraint: $3q_1 <= 12$.

We transform the inequation into an equation to define the straight line bordering the half-plane, i.e. $3q_1 = 12$.

This means that $q_1 = 4$.

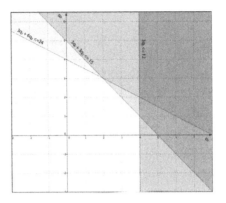

Figure 8.3. *Plot of the 3rd half-plane defined by the constraint $3q_1 <= 12$. The straight line passes through $q_1=4$ regardless of the value of q_2*

Constraint: $q_1 >= 0$.

We transform the inequation into an equation to define the straight line bordering the half-plane, i.e. $q_1 = 0$.

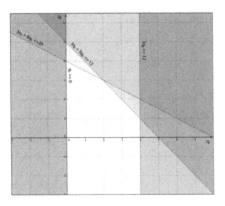

Figure 8.4. *Plot of the 4th half-plane defined by the constraint $q_1 >= 0$. The straight line passes through $q_1=0$ regardless of the value of q_2*

Constraint: $q_2 >= 0$.

We transform the inequation into an equation to define the straight line bordering the half-plane, i.e. $q_2 = 0$.

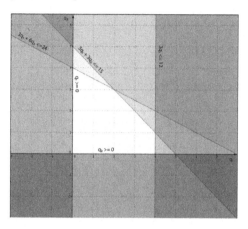

Figure 8.5. *Plot of the 5th half-plane defined by the constraint $q_2 >= 0$. The straight line passes through $q_2=0$ regardless of the value of q_1*

When all the constraints have been plotted, we find a surface representing the space of the solutions.

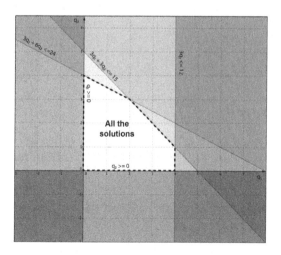

Figure 8.6. *The space of the solutions*

8.2.3. *Solution*

Once the set of the solutions has been determined, we can plot the function b of the profit we need to maximize.

We first plot the straight line for $b = 0$;

$$6q_1 + 4q_2 = 0 => 3q_1 + 2q_2 = 0$$

We calculate two points of this straight line such as to be able to plot it:

With $q_1 = -2$, we have: $q_2 = \dfrac{-3 \times -2}{2} = \dfrac{6}{2} = 3$

With $q_1 = 2$, we have: $q_2 = \dfrac{-3 \times 2}{2} = \dfrac{-6}{2} = -3$

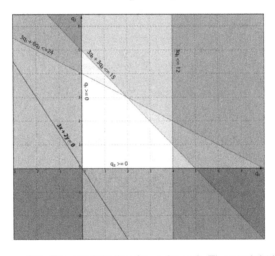

Figure 8.7. *The straight line $3q_1 + 2q_2 = 0$. The straight line passes through the pair of coordinates (-2, 2) and (2, -3)*

This straight line is in the form $f(x) = ax + b$ (**affine function** is $x = q1$ and $y = q2$).

It has a leading coefficient of -3/2 which we can easily find out by expressing q_2 according to q_1, i.e.:

$$3q_1 + 2q_2 = 0 => q_2 = -\frac{3}{2}q_1$$

COMMENT 8.2.– Here b = 0. We therefore have an unusual case of linear function: (f(x) = ax, straight line passing through the origin).

We will now transfer this straight line – its leading coefficient remains fixed – so that it offers the maximum profit, i.e. that it cuts the space of the solution in its optimum i.e. the point (*4, 1*), which can be obtained with *b = 28*.

We plot the straight line for *b = 28*.

$$6q_1 + 4q_2 = 28 => 3q_1 + 2q_2 = 14$$

We calculate 2 points of this straight line such as to be able to plot it:

With q_1 *= 2, we have:* $q_2 = \dfrac{-3\times2+14}{2} = \dfrac{8}{2} = 4$

With q_1 *= 4, we have:* $q_2 = \dfrac{-3\times4+14}{2} = \dfrac{2}{2} = 1$

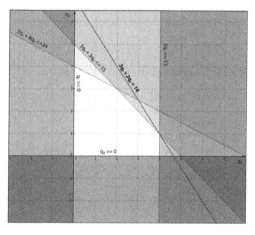

Figure 8.8. *The solution showing the optimum obtained for b = 28.*
The straight line passes through the pair of coordinates (2, 4) and (4, 1)

We can deduce:

– the hourly quantity to be produced for product P1, q_1 = 4;

– the hourly quantity to be produced for product P2, q_2 = 1;

– maximum profit B = $3q_1 + 2q_2$ = 14 => 3 × 4 + 2 × 1 = \$14.00.

To conclude, it can be seen that this method is easy to use with 2 variables (work in the plane: 2 axes). Nevertheless, it is considerably more complicated with 3 variables as we need to work with a frame of reference with 3 axes (in the space). It is impossible to work with more variables.

8.3. Simplex method

The simplex algorithm was created by George Dantzig in 1947. It is a general method for solving linear programming problems.

This technique reaches the optimal solution via successive phases. At each phase, we calculate the economic value of a solution. As the number of potential solutions is infinite, simplex only explores a limited number among which is the optimum.

8.3.1. *Method*

The algorithm can be broken down into 6 main steps:

– formalizing the problem;

– changing from the **canonical form** into the **standard form** of solution;

– creating the start table according to the equation system;

– determining the **pivot** in the table;

– iterations according to a defined and determined sequence;

– interpreting the results.

8.3.2. *An example*

Let us consider a factory that manufactures 3 products: P1, P2 and P3.

The sale of each product P1, P2 and P3 makes $8.00, $3.50 and $6.00, respectively.

To make the products, each product requires the raw material and a number of processing operations. These are described in the table below.

Product	Mass of raw material (kg)	Processing machine time (h)	Profit ($)
P1	6	3	8
P2	4	15	3.5
P3	4	2	6

Table 8.2. *The set of data for the example being processed*

Each week we have 10 tons of raw material and 6000 h of machine time. The warehouses' capacity to store the products is limited to 3500 units per week for all products.

The problem consists of defining which number of products *P1*, *P2* and *P3* should be produced per week to get maximum profit.

8.3.3. *Formalization*

In this step, we will define the variables and the constraints to be used to solve our problem.

The quantities of products P1, P2 and P3 manufactured each week will be represented by x_1, x_2 and x_3 and the profit z.

The constraints we have found after analyzing the problem are:

– processing machine time (in hours): $3x_1 + 1.5x_2 + 2x_3 \leq 6000$;

– mass of raw material (in kg): $6x_1 + 4x_2 + 4x_3 \leq 10000$;

– storage capacity (in units): $x_1 + x_2 + x_3 \leq 3500$;

with x_1, x_2 and $x_3 \geq 0$.

This gives us a system of 3 linear inequations (in the form $a_1x_1 + a_2x_2 + a_3x_3 + \dots + a_nx_n \leq b$) with 3 unknowns.

$$\begin{cases} 3x_1 + 1.5x_2 + 2x_3 \leq 6000 \\ 6x_1 + 4x_2 + 4x_3 \leq 10000 \\ x_1 + x_2 + x_3 \leq 3500 \end{cases} \qquad [8.1]$$

COMMENT 8.3.– While looking for the constraints, an expression in the form $a_1x_1 + a_2x_2 + a_3x_3 + \dots + a_nx_n \leq b$ can appear. However, by multiplying the two terms of the inequation by a negative number, we change the direction of the inequality, thus meaning that we can always express an inequation in the form $a_1x_1 + a_2x_2 + a_3x_3 + ... + a_nx_n \leq b$ which is suited to the simplex method.

The profit made each week is: $z = 8x_1 + 3,5x_2 + 6x_3$.

Known as the **objective function** (or **economic function**), this expression must be maximized or minimized.

8.3.4. *Change into standard form*

The canonical form is a form of linear programming where all the constraints are inequalities or the variables are obliged to be positive.

The change into standard form consists of transforming the inequality constraints into equality constraints by making the variables positive.

To do so, we will add a **slack variable** (called an **artificial variable**), to each constraint, which makes them positive.

Indeed, if we have $a_1x_1 + a_2x_2 + a_3x_3 + ... + a_nx_n \leq b$ and we add to the first member of the inequation $x_n + 1$, once we have verified that $x_n + 1 \geq 0$, we get $a_1x_1 + a_2x_2 + a_3x_3 + ... + a_nx_n + x_n + 1 = b$.

By asking: $x_n + 1 = b - (a_1x_1 + a_2x_2 + a_3x_3 + ... + a_nx_n)$, it can be seen that we have a slack.

Let us take e_1, e_2 and e_3 as slack variables:

– e_1 represents the number of remaining processing time (slack);

– e_2 represents the remaining storage capacity (slack);

– e_3 represents the remaining mass of material (slack).

From inequation system [8.1], we obtain an equation system in standard form:

$$\begin{cases} 3x_1 + 1.5x_2 + 2x_3 + e_1 = 6000 \\ 6x_1 + 4x_2 + 4x_3 + e_2 = 10000 \\ x_1 + x_2 + x_3 + e_3 = 3500 \end{cases} \quad [8.2]$$

where

$$x_1, x_2, x_3, e_1, e_2 \text{ and } e_3 \geq 0$$

and its cost function $z = 8x_1 + 3.5x_2 + 6x_3$.

8.3.5. *Creating the table*

Starting with system [8.2], we can build the table below, which enables us to apply the simplex algorithm.

	x_1	x_2	x_3	e_1	e_2	e_3	b_i	r
e_1	3	1.5	2	1	0	0	6000	
e_2	6	4	4	0	1	0	10,000	
e_3	1	1	1	0	0	1	3500	
z	8	3.5	6	0	0	0		

Table 8.3. *The simplex table*

Different data is included in this table:

– the columns x, y and z which represent the **basic variables**;

– the columns $e1$, $e2$ and $e3$ which represent the **non-basic variables**;

– the column r that we call the minimum ratio test (**MRT**);

– the line z, which contains the coefficients of objective function in the columns x_1, x_2 and x_3;

– the cells at the intersection of column z and b_i which contain the result of the objective function.

8.3.6. *Determining the pivot*

We now look for the pivot. There are 4 operations to do so:

1) Look for the highest column or cost coefficient.

2) Calculate the relations $r = b_i$ / *Value of the cell of the previously found column.*

3) Determine the line containing the lowest positive value of r.

4) The pivot is found at the intersection of the column found in 1 and the line found in 3.

The highest cost coefficient is $z = 8$ in the column x_1

$$r_{e1} = 6000/3 = 2000$$

$$r_{e2} = 10,000/6 = 1666.667$$

$$r_{e3} = 3500/1 = 3500$$

For our problem, the pivot is 6 (intersection of the line e_2 and the column x_1).

	x_1	x_2	x_3	e_1	e_2	e_3	b_i	r
e_1	3	1.5	2	1	0	0	6000	2000
e_2	6	4	4	0	1	0	10,000	1666.67
e_3	1	1	1	0	0	1	3500	3500
z	8	3.5	6	0	0	0		

Table 8.4. *Determining the pivot*

We can now consider the variable x_1 to be a resource. We say that x_1 "enters the base" (x_1 is an entering variable) and that e_2 "leaves the base" (e_2 is a departing variable).

	x_1	x_2	x_3	e_1	e_2	e_3	b_i	r
e_1	3	1.5	2	1	0	0	6000	2000
x_1	6	4	4	0	1	0	10,000	1666.67
e_3	1	1	1	0	0	1	3500	3500
z	8	3.5	6	0	0	0		

Table 8.5. *Entering and leaving the base*

8.3.7. *Iterations*

Calculating each of the values contained in the cells of the table at each iteration is simple and repetitive.

Two cases can occur:

1) we are on the line of the pivot;

2) we are not on the line of the pivot.

Case no. 1: the result is equal to the content of the cell divided by the value of the pivot.

Case no. 2: the result is equal to the content of the cell minus the value of the element corresponding to the line of the pivot, multiplied by the value of the element corresponding to the column of the pivot, divided by the value of the pivot.

In this new table, to move onto the next iteration, we look for the pivot again using the same principle.

The iterative loop is finished when all the coefficients are negative or null.

8.3.7.1. *Iteration no. 1 of our problem*

We apply these calculations to each of the lines in our table.

Line e_1	
Column x_1	$3 - (6 \times 3/6) = 0$
Column x_2	$1.5 - (4 \times 3/6) = -0.5$
Column x_3	$2 - (4 \times 3/6) = 0$
Column e_1	$1 - (0 \times 3/6) = 1$
Column e_2	$0 - (1 \times 3/6) = -0.5$
Column e_3	$0 - (0 \times 3/6) = 0$
Column b_i	$6000 - (10{,}000 \times 3/6) = 1000$

Table 8.6. *Calculation of the line e_1*

Line e_2 (Line of the pivot)	
Column x_1	$6/6 = 1$
Column x_2	$4/6 = 2/3$
Column x_3	$4/6 = 2/3$
Column e_1	$0/6 = 0$
Column e_2	$1/6$
Column e_3	$0/6 = 0$
Column b_i	$10{,}000/6 = 5000/3$

Table 8.7. *Calculation of the line e_2*

	Line e_3
Column x_1	$1 - (6 \times 1/6) = 0$
Column x_2	$1 - (4 \times 1/6) = 1/3$
Column x_3	$1 - (4 \times 1/6) = 1/3$
Column e_1	$0 - (0 \times 1/6) = 0$
Column e_2	$0 - (1 \times 1/6) = -1/6$
Column e_3	$1 - (0 \times 1/6) = 1$
Column b_i	$3500 - (10,000 \times 1/6) = 5500/3$

Table 8.8. *Calculation of the line e_3*

	Line z
Column x_1	$8 - (6 \times 8/6) = 0$
Column x_2	$3.5 - (4 \times 8/6) = -11/6$
Column x_3	$6 - (4 \times 8/6) = 2/3$
Column e_1	$0 - (0 \times 8/6) = 0$
Column e_2	$0 - (1 \times 8/6) = -4/3$
Column e_3	$0 - (0 \times 8/6) = 0$
Column b_i	$0 - (10,000 \times 8/6) = -40,000/3$

Table 8.9. *Calculation of the line z*

The results are reported (see Table 8.10).

	x_1	v	x_3	e_1	e_2	e_3	b_i	r
e_1	0	-0.5	0	1	-0.5	0	1000	
x_1	1	2/3	2/3	0	1/6	0	5000/3	
e_3	0	1/3	1/3	0	-1/6	1	5500/3	
z	0	-11/6	2/3	0	-4/3	0	-40,000/3	

Table 8.10. *The new table at iteration no. 1*

8.3.7.2. *Iteration no. 2 of our problem*

	x_1	x_2	x_3	e_1	e_2	e_3	b_i	r
e_1	0	-0.5	0	1	-0.5	0	1000	∞
x_1	1	2/3	2/3	0	1/6	0	5000/3	2500
e_3	0	1/3	1/3	0	-1/6	1	5500/3	5500
z	0	-11/6	2/3	0	-4/3	0	-40,000/3	

Table 8.11. *Determining the new pivot*

The pivot is *2/3* (intersection of line x_1 and column x_3).

The variable x_3 becomes a resource, x_3 "enters the base" and x_1 "leaves the base".

Once the calculations for each line of the table have been applied, like for iteration no. 1, we get the table below.

	x_1	x_2	x_3	e_1	e_2	e_3	b_i	r
e_1	0	-0.5	0	1	-0.5	0	1000	
x_3	1.5	1	1	0	0.25	0	2500	
e_3	-0.5	0	0	0	-0.25	1	1000	
z	-1	-2.5	0	0	-1.5	0	-15,000	

Table 8.12. *The new table at iteration no. 2*

It can be stated that all the cost coefficients have become negative or null and we can finish at this iteration.

8.3.8. Interpretation

The maximum profit is \$15,000 for a production of 2500 kg of product P3 (x_3). This can be verified by replacing x_1, x_2 and x_3 by their values in the cost function $z = 8x_1 + 3,5x_2 + 6x_3$, i.e.

$$z = 8 \times 0 + 3.5 \times 0 + 6 \times 2500 = 15000$$

COMMENTS 8.4.–

When one or a number of slack variables remain in the base when all the iterations have finished, it is not saturated and has an unconsumed remainder.

During maximization, the coefficients of the cost function are all negative, the basic variables are null and the non-basic variables are negative.

During minimization, the coefficients of the cost function are all positive, the basic variables are null and the non-basic variables are positive.

8.4. Duality

By default, when considering a linear program, it is said to be a **primal problem**. It is always associated with another linear program called a **dual problem**.

It is said, through a misuse of language, that these programs are symmetrical. Indeed:

– each primal constraint has a corresponding dual variable and each primal variable has a corresponding dual constraint;

– the second members of the constraints of the dual problem are the cost coefficients of the primal variables;

– the cost coefficients of the dual variables are the values present in the second member of the primal constraints;

– the constraints of type ≤ correspond to the constraints of type ≥ and vice versa;

– the optimization direction of the dual problem is the opposite of the primal problem.

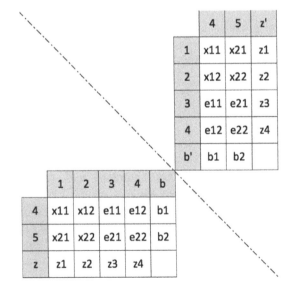

	4	5	z'
1	x11	x21	z1
2	x12	x22	z2
3	e11	e21	z3
4	e12	e22	z4
b'	b1	b2	

	1	2	3	4	b
4	x11	x12	e11	e12	b1
5	x21	x22	e21	e22	b2
z	z1	z2	z3	z4	

Figure 8.9. *The symmetry also exists between the 2 simplex tables (primal bottom left, dual top right)*

COMMENT 8.5.– It should be noted that the dual of the dual is the primal.

8.4.1. Dual formulation

The dual can only exist if the primal has been put in standard form.

Let there be three variables c, b and A

$$c \in \mathbb{R}^n, b \in \mathbb{R}^m \ and \ A \in \mathbb{R}^{n \times m}$$

$$min \sum_{i=1}^m b_i y_i \ \ with \ min \ \ b^T y$$

under constraints:

$$\sum_{i=1}^m a_{ij} y_i \geq c_j, j = 1, ..., n \ \ with \ A^T y \geq c \ and \ y \geq 0$$

$$y_i \geq 0, i = 1, ..., m.$$

8.4.2. Moving from the primal to the dual, formalization

Let us return to the example described in section 8.3.2 to show the move into 2 linear programs.

Primal constraints and objective:

$$\begin{cases} 3x_1 + 1.5x_2 + 2x_3 \leq 6000 \\ 6x_1 + 4x_2 + 4x_3 \leq 10000 \\ \ \ x_1 + x_2 + x_3 \leq 3500 \end{cases}$$

Max with $z = 8x_1 + 3.5x_2 + 6x_3$.

We set these constraints out in a table to explain the rules that will transfer the primal to the dual.

y_1	$3x_1$	$1.5x_2$	$2x_3$	6000
y_2	$6x_1$	$4x_2$	$4x_3$	10,000
y_3	x_1	x_2	x_3	3500
Will become dual	1st dual constraint	2nd dual constraint	3rd dual constraint	Objective function

Table 8.13. *From the primal to the dual*

Cost coefficients of the primal:

z	$8x_1$	$3.5x_2$	$6x_3$	Will become second member

Table 8.14. *Cost coefficients of the primal*

By applying these rules, we obtain:

Dual constraints and objective:

$$\begin{cases} 3y_1 + 6y_2 + y_3 \geq 8 \\ 1.5y_1 + 4y_2 + y_3 \geq 3.5 \\ 2y_1 + 4y_2 + y_3 \geq 6 \end{cases}$$

Min with z' = $6000y_1 + 10000y_2 + 3500y_3$

To change the direction of the inequalities such as to correctly apply the simplex, we will multiply the two members by -1, i.e.:

$$\begin{cases} -3y_1 - 6y_2 - y_3 \leq -8 \\ -1.5y_1 - 4y_2 - y_3 \leq -3.5 \\ -2y_1 - 4y_2 - y_3 \leq -6 \end{cases}$$

We must not forget also to inverse the sign of each of the coefficients of the cost function *z'*.

The simplex table becomes:

	y_1	y_2	y_3	e'_1	e'_2	e'_3	b'_i
e'_1	-3	-6	-1	1	0	0	-8
e'_2	-1.5	-4	-1	0	1	0	-3.5
e'_3	-2	-4	-1	0	0	1	-6
z'	-6000	-10,000	-3500	0	0	0	
r'							

Table 8.15. *Dual simplex table*

8.4.3. *Determining the pivot*

In the dual, determining the pivot also follows, if it can be said, a symmetry:

1) looking for the line where b'_i is lowest;

2) calculating the relations r' = z' / Value of the cell of the previously found line;

3) determining the column containing the lowest positive value of r';

4) the pivot is found at the intersection of the line found at 1 and of the column found at 3.

In our case, the pivot is *-6* (intersection of the line e'_1 and the column y_2).

We apply the algorithm and obtain:

	y_1	y_2	y_3	e'_1	e'_2	e'_3	b'_i
e'_1	-3	-6	-1	1	0	0	-8
e'_2	-1.5	-4	-1	0	1	0	-3.5
e'_3	-2	-4	-1	0	0	1	-6
z'	-6000	-10,000	-3500	0	0	0	
r'	2000	5000/6	3500	0	∞	∞	

Table 8.16. *Determining the pivot*

The variable y_2 becomes a resource, y_2 "enters the base" and e'_1 "leaves the base".

8.4.4. *Iterations*

Like for the primal, we will roll out iterations by respecting the same calculation method (see section 8.3.6).

8.4.4.1. *Iteration no. 1*

	y_1	y_2	y_3	e'_1	e'_2	e'_3	b'_i
y_2	0.5	1	1/6	-1/6	0	0	4/3
e'_2	0.5	0	-1/3	-2/3	1	0	11/6
e'_3	0	0	-1/3	-2/3	0	1	-2/3
z'	-1000	0	-5500/3	-5000/3	0	0	40,000/3
r'	-2000	0	-11,000	10,000	∞	∞	

Table 8.17. *The simplex at iteration no. 1*

The pivot is *-2/3* (line e'_3, column e'_1).

The variable e'_1 becomes a resource, e'_1 "enters the base" and e'_3 "leaves the base".

8.4.4.2. Iteration no. 2

	y_1	y_2	y_3	e'_1	e'_2	e'_3	b'_i
y_2	0.5	1	1/4	0	0	-1/4	3/2
e'_2	0.5	0	0	0	1	-1	5/2
e'_1	0	0	1/2	1	0	-3/2	1
z'	-1000	0	-1000	0	0	-2500	15,000
r'							

Table 8.18. *The simplex at iteration no. 2*

The variable e'_1 becomes a resource, e'_1 "enters the base" and e'_3 "leaves the base".

It can be stated that all the cost coefficients have become negative or null and we can therefore finish at this iteration.

8.4.5. Interpretation

The maximum profit is \$15,000. This can be verified by replacing y_1, y_2 and y_3 with their values in the cost function $z' = 6000y_1 + 10000y_2 + 3500y_3$, i.e.:

$$z = 6000 \times 0 + 10000 \times 3/2 + 3500 \times 0 = 15000$$

COMMENT 8.6.– If the primal has lots of constraints and fewer variables, the simplex will be more effective with the dual.

8.5. Exercises

8.5.1. Exercise 1: Video and festival

The Chalon sur Saône *arts de la rue* festival takes place each year in July. Paul is unable to attend and asks his friend Jacques, who lives in the town, to video him a several shows. There are a large number of shows taking place one after the other throughout the day from 10 AM to 11.30 PM. Excluding the lunch break and the various intervals, there is a total of 9.5 h of shows to be filmed.

Jacques' camera films in HD and stores what it records on SD memory cards.

Jacques can only get his hands on 8 SD 8Go cards that can store 65 min of HD video each and 3 SD 16Go cards that can store 130 min.

The SD 8Go cards costs $10.00 each and the 16 Go $15.00 each.

Jacques wants to know how many SD 8Go and 16Go cards he should buy to be able to film all of the shows while spending the minimum amount.

Question:

Graphically determine the number of SD 8 and 16Go cards that Jacques should buy.

8.5.2. *Exercise 2: Simplex*

Let us consider the following optimization problem:

$$\max z = 5x + 6y$$

under the constraints:

$$3x + 2y \leq 10$$

$$9x + 4y \leq 36$$

$$x + 2y \leq 8$$

Questions:

1) Solve the problem above using the graphic method.

2) Confirm the result with the simplex.

8.5.3. *Exercise 3: Primal and dual*

Let us consider the following optimization problem:

$$\max z = 20x1 + 30x2$$

under the constraints:

$$2x1 + x2 \leq 1000$$

$$x1 + x2 \leq 800$$

$$x1 \leq 400$$

$$x2 \leq 700$$

Questions:

1) Solve, using the simplex method, the primal linear program attached to the above problem.

2) Confirm the result by applying duality.

8.5.4. *Solution to exercise 1*

The variables x and y will designate the number of SD 8Go and SD 16Go cards, respectively.

Recording should last *9.5 h* maximum, i.e. *570 min*. To cover all the shows, Jacques must have at least *570 mins* of recording time on the cards.

We can therefore define the following constraints:

– number of SD 8Go cards: $x \leq 6$;

– number of SD 16Go cards: $y \leq 3$;

– available recording time: *65x + 130y ≥ 570* ⇔ *13x + 26y ≥ 114*;

– x and y must be positive integers (we cannot divide an SD card) i.e. $x >$ *0* and $y >$ *0*.

We plot all the equations in a framework to determine all the solutions.

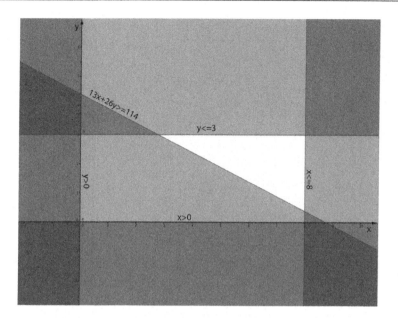

Figure 8.10. *The constraints and the cost function defining all the solutions are represented by the white section of the graph*

In this triangle, only the points with full coordinates are solutions. The borders, which are also solutions, must also be taken into account as our constraints are of lower than or equal or higher or equal constraints.

The only solutions kept – which meet the constraints – are the following 11 pairs: *(4, 3) ; (5, 3) ; (6, 3) ; (7, 3) ; (8, 3) ; (5, 2) ; (6, 2) ; (7, 2) ; (8, 2) ; (7, 1)* and *(8, 1)*.

To film all the shows (570 min) Jacques could buy:

– 4 SD 8Go cards and 3 SD 16Go cards, i.e. $4 \times 65 + 3 \times 130 = 650$ min for a cost of $4 \times 10 + 3 \times 15 = \$95.00$;

– or 8 SD 8Go cards and 1 SD 16Go card, i.e. $8 \times 65 + 1 \times 130 = 650$ min for a cost of $8 \times 10 + 1 \times 15 = \$95.00$;

– or 7 SD 8Go cards and 3 SD 16Go cards, i.e. $7 \times 65 + 3 \times 130 = 845$ min for a cost of $7 \times 10 + 3 \times 15 = \$115,00$, etc.

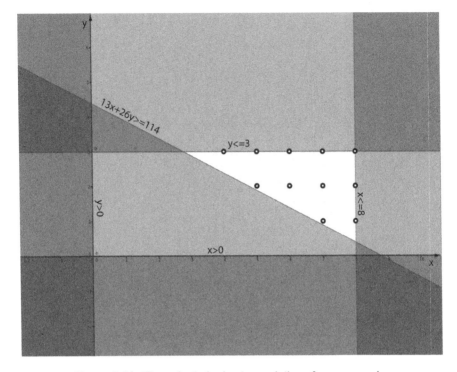

Figure 8.11. *The pairs to be kept as solutions for our exercise*

To find out the pair(s) corresponding to the minimum investment cost, we should only keep the coordinate points close to the straight line linked to the cost function, i.e.:

– pair (4, 3) whose cost is *4 ×10 + 3 ×15 = $85.00*;

– pair (5, 2) whose cost is *5 ×10 + 2 ×15 = $80.00*;

– pair (7, 1) whose cost is *7 ×10 + 1 ×15 = $85.00*.

The least expensive solution is therefore the pair *(5, 2)*. Jacques should therefore buy 5 SD 8Go cards and 2 SD 16Go cards to spend a minimum amount while being able to film all the shows (*5 ×65 + 2 ×130 = 585 min)*. He will even have an extra remaining *585 – 570 = 15 min.*

8.5.5. *Solution to exercise 2*

Question 1:

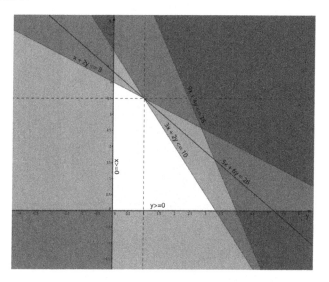

Figure 8.12. *The graphic solution*

The maximum is 26. It is obtained for x = 1 and y = 3.5.

Indeed: $5 \times 1 + 6 \times 3.5 = 26$.

Question 2:

Here is the simplex table and its 2 successive iterations.

	x	y	e1	e2	e3	Total	r
e1	3	2	1	0	0	10	5
e2	9	4	0	1	0	36	9
e3	1	2	0	0	1	8	4
	5	6	0	0	0	0	

	x	y	e1	e2	e3	Total	r
e1	2	0	1	0	-1	2	1
e2	7	0	0	1	-2	20	2.86
y	0.5	1	0	0	0.5	4	8
	2	0	0	0	-3	-24	

	x	y	e1	e2	e3	Total	r
x	1	0	0.5	0	-0.5	1	
e2	0	0	-3.5	1	1.5	13	
y	0	1	-0.25	0	0.75	3.5	
	0	0	-1	0	-2	-26	

Table 8.19. *The solution via simplex*

It can be seen in the last table that x equals 1, y is equal to 3.5 and that the maximum is worth 26. This confirms the graphic solution found in question 1.

8.5.6. Solution to exercise 3

Question 1:

Iterations of the primal.

	x_1	x_2	e_1	e_2	e_2	e_4	b_j	r
e_1	2	1	1	0	0	0	1000	1000
e_2	1	1	0	1	0	0	800	800
e_2	1	0	0	0	1	0	400	∞
e_4	0	1	0	0	0	1	700	700
z	20	30	0	0	0	0		

	x_1	x_2	e_1	e_2	e_2	e_4	b_j	r
e_1	2	0	1	0	0	-1	300	150
e_2	1	0	0	1	0	-1	100	100
e_2	1	0	0	0	1	0	400	400
x_2	0	1	0	0	0	1	700	∞
z	20	0	0	0	0	-30	-21000	

	x_1	x_2	e_1	e_2	e_2	e_4	b_j	r
e_1	0	0	1	-2	0	1	100	
x_1	1	0	0	1	0	-1	100	
e_2	0	0	0	-1	1	1	300	
x_2	0	1	0	0	0	1	700	
z	0	0	0	-20	0	-10	-23000	

Table 8.20. *The succession of iterations to calculate the simplex of the PL primal*

$$z = 20 \times 100 + 30 \times 700 = 23000$$

Question 2:

Formalization:

$$\min z' = 1000y_1 + 800y_2 + 400y_3 + 700y_4$$

$$2y_1 + y_2 + y_3 \geq 20$$

$$y_1 + y_2 + y_4 \geq 30$$

Iterations of the dual.

	y_1	y_2	y_3	y_4	e'_1	e'_2	b'_i
e'_1	-2	-1	-1	0	1	0	-20
e'_2	-1	-1	0	-1	0	1	-30
z'	-1000	-800	-400	-700	0	0	
r'	1000	800	∞	700	∞	0	

	y_1	y_2	y_3	y_4	e'_1	e'_2	b'_i
e'_1	-2	-1	-1	0	1	0	-20
y_4	1	1	0	1	0	-1	30
z'	-300	-100	-400	0	0	-700	21000
r'	150	100	400	∞	0	∞	

	y_1	y_2	y_3	y_4	e'_1	e'_2	b'_i
y_2	2	1	1	0	-1	0	20
y_4	-1	0	-1	1	1	-1	10
z'	-100	0	-300	0	-100	-700	23000
r'							

Table 8.21. *The succession of iterations to calculate the simplex of the PL dual*

$$z' = 1000 \times 0 + 800 \times 20 + 400 \times 0 + 700 \times 10 = 23000$$

9

Software

9.1. Software for OR and logistics

Earlier in this book, we presented a group of techniques for resolving numerous problems in operational research that can frequently be used in the domain of logistics.

Most of the algorithms that we have explained are often difficult to implement and require a great deal of time and effort to reach a satisfactory solution.

The examples and exercises in the previous chapters have been chosen with care but do not always correspond to situations that might be found in business or on the ground. In most instances, they have been scaled down to show you the development and application of simple calculations.

In reality, and although it is important to know the fundamental bases for resolving this or that problem, using a computer is indispensable. Its functions, power and omnipresence provide the user with a comfort, ease of use and ergonomics that it is difficult to surpass. In a very short time, it can adapt to all situations and can resolve the most difficult problems, often requiring an enormous number of iterations and calculations, provided that the user learns to operate it, along with some software functions.

In this chapter, we are going to introduce the main tools for tackling different logistical problems that might be encountered in a professional setting.

It is clear that we do not aim to be able to show all the existing products and all their functions, this work is not long enough and that is not its goal. Above all, we wish to introduce you to some formulae that you can improve upon with experience.

In this first chapter of the second part, we will limit ourselves to introducing you to some market software that has been adapted or is adaptable for resolving common problems.

9.2. Spreadsheets

We must go back to 1961, to Richard Mattessich, to find the first spreadsheet or rather the first program for manipulating accounting worksheets. In 1969, Rene Pardo and Remy Landau invented "LANPAR" (LANguage for Programming Arrays at random) a language for working on computerized accounting spreadsheets. It would be used by Bell Canada, AT&T and General Motors to carry out changes in their accounting forms. To this end, an American patent was placed (US patent number 4,398,249).

However, the real development of spreadsheets as they are known today is due to Dan Bricklin who in 1978 created an accounting worksheet capable of handling a matrix of 20 columns and five lines. To improve it, Bricklin would be joined by Bob Frankston, himself later joined by Daniel Fylstra. In May 1979, the software was first marketed under the name of "Visicalc" (visible calculator). It was a great success and around 1 million copies would be sold.

It would be followed by Mitch Kapor's spreadsheet Lotus 1-2-3 in 1982, then Microsoft's Multiplan, also in 1982, then Microsoft Excel in 1984, and many others.

Today, spreadsheets have become indispensable and scientists and managers could no longer do without them. It is an extraordinary tool for anyone who carries out or creates complex and repetitive calculations, simulations and dashboards.

The name "spreadsheet" is sometimes replaced by "electronic accounting worksheet" or "electronic spreadsheet", it is in fact an enormous matrix formed of lines and columns that define the cells.

Each of these cells can contain text, a number, a date, a time, a simple mathematical formula or even other objects such as images, logos, etc.

Because of its structure, a spreadsheet can group several matrices or sheets that can exchange information and communicate with each other in a single file. The files themselves can also exchange data.

Spreadsheets today have been joined by numerous other tools to create graphics, carry out cross-analysis, manage scenarios, exchange data with databases, etc.

Numerous types of software are available on the market. Most are multiplatform and function with Microsoft Windows, Mac OSX, Linux, iOS, Android or with a thin client, that is to say via an Internet connection and navigation software (Firefox, Internet Explorer, Safari, Chrome, etc.).

In this work, we will use the spreadsheet Microsoft Excel with its advanced functions, **pivot table** management and programming language, **Visual Basic for Applications (VBAs)**. The examples shown can, however, be adapted for others.

Here is a small, non-exhaustive list of spreadsheets available on several platforms. Many form part of an office suite.

Microsoft Windows:

– Microsoft Excel (taken from Microsoft Office);

– LibreCalc (taken from the suite LibreOffice – free);

– Calc (taken from the suite OpenOffice – free).

Mac OSx:

– Apple iWork Numbers;

– Microsoft Excel;

– Calc (taken from the suite OpenOffice – free);

– Calc (taken from the suite NeoOffice).

Android:

– Andropen Office (Open Office android – free);

– Smart Office;

– Office Suite Pro;

– Quick Office (Google office suite– free).

iOS:

– QuickOffice;

– Office2 HD;

– Calc XLS Spreadsheet free (free);

– Calc XLS Spreadsheet;

– Spreadsheet 365 (free);

– Spreadsheet – Excel edition HD.

Online tools (accessible via a web browser):

– Google Sheets (Google suite);

– Zoho Sheet (Zoho suite);

– Microsoft Excel OnLine (Microsoft suite).

9.2.1. *Advanced functions with Microsoft Excel*

We are now going to explain some of the principles and advanced functions that we will make use of in later exercises.

9.2.1.1. *Calendars and dates*

For Microsoft Excel, all dates are integers and hours are decimal fractions. This presents a significant advantage since on this principle it is possible to add, subtract or compare dates or hours as if they were simple numbers.

The number 1 or in the case of a date, the **series number** 1 corresponds to January 1st 1900 at 0h 0min 0sec (01/01/1900 at 12:00:00 AM).

The hours are decimal numbers comprising between 0.0 for 00:00:00 and 0.99999 for 23:59:59.

By adding an integer part and a decimal part, it is possible to describe a date and an hour, for example:

42000.5 = 27/12/2014 12:00:00

42019.251 = 15/01/2015 6:01:26

9.2.1.2. Conditions associated with SUM and COUNT

The conditional branching if...then can be linked to the functions SUM or COUNT with (SUMIF and COUNTIF) or several criteria (SUMIFS or COUNTIFS).

The principle is always the same, the main function is carried out if the conditions are met.

SUMIF(C2:C13;"M";F2:F13) will result in:

If the cells from C2 to C13 contain the letter "M", add the corresponding content from F2 to F13.

SUMIFS(F2:F13;C2:C13;"M";D2:D13;">35") will result in:

If the cells from C2 to C13 contain the letter "M" and cells D2 to D13 contain a number greater than 35, add the corresponding contents from F2 to F13.

	A	B	C	D	E	F	G
1	Name	First name	Gender	Age	Job	Salary	
2	WILSON	John	M	32	Engineer	$3 800,00	
3	PIERCE	Mary	F	25	Engineer	$3 500,00	
4	SANDERS	James	M	34	Accountant	$2 000,00	
5	SMITH	Harold	M	33	Marketing	$2 200,00	
6	BENTON	Cindy	F	26	Secretary	$2 200,00	
7	MARTIN	Jane	F	25	Assistant	$2 600,00	
8	CARSON	Kate	F	29	Secretary	$2 200,00	
9	DUMOULIN	Barclay	M	41	Logistician	$2 450,00	
10	WALSH	Mark	M	36	Marketing	$2 500,00	
11	DELAHAYE	Colin	M	38	Marketing	$2 500,00	
12	CODD	Peter	M	21	Technician	$2 600,00	
13	RICHMOND	Emily	F	24	Technician	$2 100,00	
14							
15	Sum total of salaries for male employees					**$18 050,00**	
16	Sum total of salaries for male employees aged over 35					**$7 450,00**	
17							

Figure 9.1. *Cell F15 contains the formula: =SUMIF(C2:D13; "M";F2:F13) which calculates the sum total of the salaries for male employees. Cell F16 contains the formula: =SUMIFS(F2:F13;C2:C13;"M" ;D2:D13;">35") which calculates the sum total of salaries for male employees aged over 35*

9.2.1.3. *Vertical and horizontal searches*

Two very useful and much used functions of Microsoft Excel are VLOOKUP and HLOOKUP, for vertical and horizontal searches, respectively.

Their syntaxes are the following:

– VLOOKUP ("Value sought"; "Range of cells where sought; "No. of the column containing the data to be displayed"; "Type of value sought");

– HLOOKUP ("Value sought"; "Range of cells where sought"; "No. of the line containing the data to be displayed"; "Type of value sought").

Here is an example applied on the table in Figure 9.2.

	A	B	C	D	E	F	G	H	I
1	Name	First name	Gender	Age	Job	Salary		Name :	MARTIN
2	BENTON	Cindy	F	26	Secretary	$2 200,00		Job :	Assistant
3	CARSON	Kate	F	29	Secretary	$2 200,00			
4	CODD	Peter	M	21	Technician	$2 600,00			
5	DELAHAYE	Colin	M	38	Marketing	$2 500,00			
6	DUMOULIN	Barclay	M	41	Logistician	$2 450,00			
7	MARTIN	Jane	F	25	Assistant	$2 600,00			
8	PIERCE	Mary	F	25	Engineer	$3 500,00			
9	RICHMOND	Emily	F	24	Technician	$2 100,00			
10	SANDERS	James	M	34	Accountant	$2 000,00			
11	SMITH	Harold	M	33	Marketing	$2 200,00			
12	WALSH	Mark	M	36	Marketing	$2 500,00			
13	WILSON	John	M	32	Engineer	$3 800,00			
14									

Figure 9.2. *Table with the function VLOOKUP applied*

Knowing the name of an employee, placed in cell I1, we want to know his or her job, in I2 .

=VLOOKUP(I1;A2:E13;5;0)

This formula results in: we seek the value placed in I1, in the first column (A), within the range of cells A2 to E13, to obtain a corresponding value placed in the fifth column (E) of the range being considered.

The spreadsheet seeks "MARTIN" and finds it on line 7 of the first column (A) and then it looks on the same line, in the fifth column of the range (column E) and displays the content of the cell, which is "Assistant".

The function HLOOKUP operates on the same principle, but the content is sought in the line corresponding to the range considered and not in the column.

The type being equal to 0 (or FALSE), the search hits a value exactly equal to "MARTIN", otherwise with 1 (or TRUE), it would strike an approximate value (with a first, sorted column).

NOTE.– There is also a LOOKUP function, which is less powerful since it requires sorted data (increasing or decreasing) to function correctly:

LOOKUP (Value sought"; "Vector-range of cells where it is sought";" Vector-range of cells containing the result").

Here, the search range is a vector, that is to say a group of contiguous cells restricted to a line or a column. The same goes for the cells containing the result.

9.2.2. Pivot tables

A pivot table is a tool attached to a spreadsheet and used to synthesize data resulting from a worksheet. It proceeds by groupings, crossings and various types of calculations: sums, differences, products, averages, variances, counts, personalized formulae, etc. to generate an interactive multidimensional table, built along two main axes, abscissas and ordinates and perhaps specific rubrics, linked to one or more categories.

For clarification, we suggest you to look at the simple example below.

	A	B	C	D	E	F
1	Name	First name	Gender	Age	Job	Salary
2	WILSON	John	M	32	Engineer	$3 800,00
3	PIERCE	Mary	F	25	Engineer	$3 500,00
4	SANDERS	James	M	34	Accountant	$2 000,00
5	SMITH	Harold	M	33	Marketing	$2 200,00
6	BENTON	Cindy	F	26	Secretary	$2 200,00
7	MARTIN	Jane	F	25	Assistant	$2 600,00
8	CARSON	Kate	F	29	Secretary	$2 200,00
9	DUMOULIN	Barclay	M	41	Logistician	$2 450,00
10	WALSH	Mark	M	36	Marketing	$2 500,00
11	DELAHAYE	Colin	M	38	Marketing	$2 500,00
12	CODD	Peter	M	21	Technician	$2 600,00
13	RICHMOND	Emily	F	24	Technician	$2 100,00

Figure 9.3. *A database of salaries in a Microsoft Excel accounting sheet*

In Figure 9.3, we can see a table displaying a list of employees in a business mentioning their names, first names, gender, age, job and monthly salary.

To know the average salary for each category of employee in relation to their jobs and identifying their gender, we can create the pivot table in Figure 9.4.

	H	I	J	K
Average salary	Gender			
Job	F	M	Total	
Accountant		$2 000,00	$2 000,00	
Assistant	$2 600,00		$2 600,00	
Engineer	$3 500,00	$3 800,00	$3 650,00	
Logistician		$2 450,00	$2 450,00	
Marketing		$2 400,00	$2 400,00	
Secretary	$2 200,00		$2 200,00	
Technician	$2 100,00	$2 600,00	$2 350,00	
Total	**$2 520,00**	**$2 578,57**	**$2 554,17**	

Figure 9.4. *An example of a pivot table in Microsoft Excel*

At the intersection of "job" and "gender", we find the average salary. In the last line is the total average salary for all jobs combined for both sexes and in the last column, the total average salary for both sexes combined according to their jobs in the business.

Figure 9.5 shows a simple example of a pivot table created in a few clicks, numerous other functions exist and offer the chance to generate very sophisticated relationships.

In parallel with pivot tables, it is also possible to create pivot graphs.

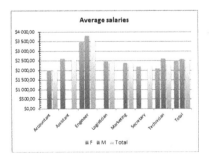

Figure 9.5. *A pivot graph resulting from the previous pivot table. For a color version of the figure, see www.iste.co.uk/reveillac/logistics.zip*

9.2.3. *Solver*

The Excel solver is a very powerful tool for finding an optimal value, called the objective, depending on constraints or limits placed in a group of cells. To function, it relies on variables known as decision variables that are included in the formulae for calculating the objective.

It is frequently used to solve numerous optimization problems such as those found in the field of operational research.

Below, you can see a small example of these possibilities, which we will expand upon in Chapter 10.

If this is the equation:

$$ax^3 + bx^2 + cx + d = 0$$

Calculate the value of the unknown *x*, with $a = 2$; $b = 5$; $c = -5$; $d = -1$

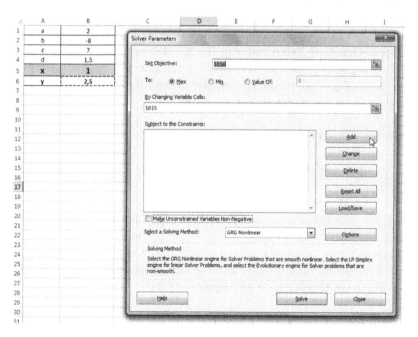

Figure 9.6. *Spreadsheet containing the variables and the solver setup window (y = 2.5 for x = 1)*

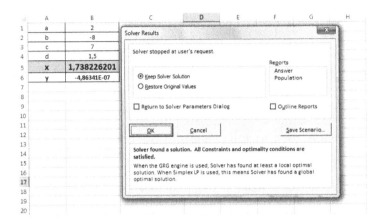

Figure 9.7. *The spreadsheet containing the variables and the result (x = 1.738226201) after launching the solver*

9.2.4. *Visual Basic for Applications*

VBA is a language used by Microsoft Office products including Excel, which enables directly executable programs to be developed. As its name indicates, it relies on implementing Basic language including the principles of **Object-Oriented Programming (OOP)**. It is simple to use although its functions are very powerful. Within the spreadsheet, it enables certain tasks to be automated and previously non-existent functions to be created, all through using the spreadsheet's capabilities.

A work such as this will not be sufficient to introduce you to VBA and to teach you to use it. We will limit ourselves to showing you some basic elements that will seem familiar to those who know the rudiments of programming.

All the examples of code that you will find in this book have a commentary in order to help them be understood. We are aware that unfortunately this will not always be sufficient for a novice and we invite you to consult the bibliography and the list of Internet links, at the end of this work, to study this language further.

9.2.4.1. *Visual Basic Editor (VBE)*

To handle VBA in the most ideal conditions, an editor is available. It offers numerous possibilities. It is above all an entry tool, but it is also an instrument for checking, structuring, fine-tuning and debugging code as well as for the conception and design of user interfaces.

Figure 9.8 shows the different elements available in VBE:

1) tool bars for editing, debugging, executing, shaping, etc.;

2) a project manager for handling workbooks and spreadsheets that are open and in use, for creating and modifying VBA modules, forms and classes;

3) a property manager provides access to different elements' characteristics;

4) a procedure editing zone for writing different VBA procedures and instructions;

5) a **class** editing zone for creating reusable objects within the code following the principles of object-oriented programming adapted for VBA;

6) a form editing zone for creating interactive windows;

7) a tool box for creating the different elements present in the forms.

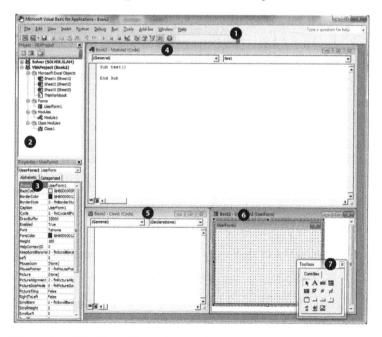

Figure 9.8. *The Visual Basic for Applications editor in Microsoft Excel*

9.2.4.2. *Variables*

In VBA, as in any other language, the variables store the data to be handled.

There are several **types** that can be defined by their declarations.

Name	Type	Details	Symbol
Byte	Numeric	Whole number from 0 to 255	
Integer	Numeric	Whole number from -32768 to 32767	%
Currency	Numeric	Fixed decimal number from -922337203685477.5808 to 922337203685477.5807	@
Single	Numeric	Floating point number from -3.402823E38 to 3.402823E38	!
Long	Numeric	Numeric -2147483648 to 2147483647	&
Double	Numeric	Floating point number from -1.79769313486232E308 to 1.79769313486232E308	#
String	Text	Text	$
Date	Date	Date and time	
Boolean	Boolean	True or false	
Object	Object	Microsoft object	
Variant	All	All types of data (by default if the variable is not declared)	

Table 9.1. *The types of variables in VBA*

There are two ways of declaring variables in VBA, **explicit** and **implicit declaration**.

An explicit declaration is made using the keyword **Dim,** for example:

```
Dim var1 As Integer
Dim Number As Long
```

```
Dim Breadth As Double
```

A variable name should begin with a letter and should only contain numbers, letters or an underscore. It should be no more than 255 characters in length, and it should not be identical to a VBA keyword. The type of the variable is defined after the keyword **As**.

An implicit declaration can be made in the form:

```
Var1 = 12
Number = 150
My_name = "Smith"
```

In this case, the variable is a variant by default.

In VBA, the variable names are not case-sensitive, `Var1` is equivalent to `var1`.

9.2.4.3. Operators

In a VBA expression, the following operators can be used:

Arithmetic operators: these combine numeric expressions and return a numeric expression.

+   Addition

-    Subtraction

*   Multiplication

/   Division

\   Division

MOD Remainder from division

^   Exponent

Comparison operators: these compare two numeric expressions and return a Boolean (true or false).

=   Equal to

>   Greater than

<   Less than

>=   Greater than or equal to

<=   Less than or equal to

<>   Different to

Concatenation operators: these link two string expressions and return one strong expression.

&   concatenate of two strings.

Logical operators: these return a Boolean expression.

And, Or, Xor, Not.

9.2.4.4. *Decision-making structures*

These are numerous: If...then...else, select...case, Iif, Switch, Choose, With. Below, we have shown only those most commonly used.

In VBA, it can take two forms:

In a single line:

If condition Then instruction1 [Else instruction2]

In this case, the condition is a Boolean expression (true or false). If it is numeric, the value 0 corresponds to false and all other values correspond to true.

In block form:

```
If condition Then
     [instructions]

Else
     [instructions]

End If
```

The structure If...Then...Else can only work with two alternatives, if there are more, Select...Case... can be used

A single expression (or a variable) placed behind Select Case is tested at the beginning, and then compared with the lists of values specified by Case. As soon as a concordance is found, the corresponding instructions are carried out and the program exits the structure. If no concordance exists, the instructions from Else are carried out.

```
Select Case expression
Case List1
     [instructions]
Case List 2
     [instructions]
[Case Else
     instructions]
End Select
```

9.2.4.5. The loops

Just as for decision-making structures, there are several solutions for creating loops: For...To...Next, Do Until...Loop, Do While...Loop, While...Wend, For Each...Next. Here too, we will only show those most commonly found.

The loop For...To...Next

This is a repetitive structure generated by a computer which increments to each passage in the loop from an initial value up to a final value.

The incrementation step can be defined via the instruction Step, which decides the value of the step, positive or negative, whole or decimal. In the absence of Step, the step is equal to 1.

The instruction Next ends the loop and prepares for transfer to the next counter value.

```
For counter = beginning To end [Step]
        [instructions]
Next
```

The loop Do Until...Loop

In this structure, a set of instructions is repeated until a condition defined by the instruction Until is true. The instruction Loop ends the loop.

```
Do Until condition
[instructions]
Loop
```

The loop Do While...Loop

Here, the performance is the same as in the loop using Until, but the set of instructions is repeated so long as the condition is true.

```
Do While condition
[instructions]
Loop
```

It should be noted that the instructions Until and While can also be placed beside Loop, which enables the condition to be tested once the block of instructions has been executed.

```
Do
        [instructions]
Loop Until condition
```

```
Do
        [instructions]
Loop While condition
```

The loop While...Wend

This loop is identical to the previous one but its syntax is different.

```
While condition
        [instructions]
Wend
```

With the latest versions of VBA, it is better to use Do...Loop instead of While...Wend, which is less supple and not so well optimized.

9.2.4.6. *Some other functions*

To insert a commentary:

An apostrophe is used to place a commentary within a VBA code.

```
'This is a commentary
```

To define a procedure:

In Microsoft Excel, a macro triggers the execution of one or several procedures (Sub) written in VBA. The procedures are stored in the VBA **modules** which can be attached to a Microsoft Excel sheet or file.

The procedure is, therefore, the basic element in which the code is written. To define a procedure, it is sufficient to give it a name preceded by the instruction Sub. Specific arguments that will be marked in parentheses can eventually be joined to this procedure name. These arguments permit values generated in one procedure to be used in another.

A procedure can be **private** or **public**. When it is private, it becomes impossible to access it outside the VBA module that contains it.

When Private is not specified, a procedure is public by default.

```
Private Sub MyProcedure1 ([arguments])
    [instructions 1]

End Sub

Sub MyProcedure2 ()
    [instructions 2]

    MyProcedure1 [values, arguments]
    [instructions 3]

End Sub
```

To select a zone of cells in a spreadsheet:

```
Range ("B2").Select
```
Selects cell B2

```
Cells (2, 4).Select
```
Selects cell D2 (line 2, column 4)

```
Range ("B2:C5").Select
```
Selects the zone of cells B2 to C5

Allocating contents from a cell to a variable:

```
Var1 = Range("B2")
```
Allocates the contents of cell B2 to the variable `Var1`

```
Var2 = Selection
```
Allocates the contents of the selection (last instruction containing `.Select`) to the variable `Var2`

```
Var3 = Cells(10, 1)
```
Allocates the contents of the cell A10 (line 10, column 1) to the variable `Var3`

To send a value to a spreadsheet cell:

```
Range("B2") = 10
```
Places the value 10 in cell B2

```
Cells(2, 3) ="John"
```
Places the chain "John" in cell C2 (line 2, column 3)

```
Range("B3") = Var1
```
Places the value that contains Var1 in cell B3

To display a dialogue box:

Dialogue boxes are very useful for displaying warnings or commentary, for making selections, repeating operations or abandoning or ignoring a process. The instruction `MsgBox` is responsible for managing them.

```
MsgBox"Hello"
```

Displays a dialogue box containing "Hello".

Figure 9.9. *A dialogue box containing "Hello"*

```
If   MsgBox("Do you wish to continue?", vbYesNo,
"Confirmation requested") = vbYes then
```

[instructions 1]

```
Else
```

[instructions 2]

```
End If
```

Displays a dialogue box, with "Confirmation requested" as its title.

This box contains the message "Do you wish to continue?".

Clicking on the *Yes* button executes instruction block1. Clicking on the *No* button executes instruction block 2.

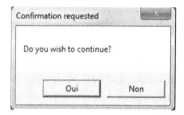

Figure 9.10. *An interactive dialogue box (Yes or No)*

Managing dialogue boxes presents numerous opportunities that are not outlined in the material presented above.

To open a window containing an input field:

A dialogue box retrieved by the instruction `InputBox` can be used to ask the user for a value.

```
Var1 = InputBox("Text", "Box title", "Default")
```

A dialogue box with the title "Box title" will be displayed and will ask the user to enter a value in the field located beneath the message "Text". By default, if the user does not enter anything, a value will be added, here, "Default".

When the user clicks on the **OK** button, the variable Var1 will be allocated either the value entered or the default value.

Clicking on the *Cancel* button cancels the input operation.

Figure 9.11. *Dialogue box ready for user input. You can see the default value in the field*

9.3. Project managers

Project management developed gradually. It did not really appear until the beginning of the 20th Century, when it was associated with different manufacturing techniques and was practiced intuitively by those who used them.

It was not until the 1950s, with the emergence of production chains and product standardization that project management became a subject of significant interest in engineering. First, it was used in large-scale projects, then in the military, automobile industry, in ship-building and architecture and then later it diversified and spread to all sectors of industrial design and manufacturing, reaching as far as finance, banking and administration.

Since the 1980s, developers have looked at the possibility of creating project management software to handle tasks and resources while calculating a program evaluation and review technique (PERT) or *méthode des potentiels métra* (MPM) network or a Gantt chart. The sheer number of calculations involved in planning and the difficulty of regularly developing them and keeping them up-to-date were the main factors leading to the creation of this type of program (recalculating graphs, networks and expenses can be very time-consuming if carried out by hand).

In 1984, Alan Boyd, responsible for development at Microsoft, was one of the first to propose and develop specifications for a powerful tool for resolving problems associated with monitoring multiple projects developed in house. He asked a firm in Seattle to develop a model that met his needs. In 1985, Microsoft repurchased the rights and would go on to market the first version of Microsoft Project, which at that time operated in MS-DOS[1].

Other software also appeared, such as Primavera Project Planner, CA Super Project, Sciforma, MicroPlanner and many others, especially after the arrival of the electronic office (Mac OS, Microsoft Windows, etc.) which brought graphic editing to the computer screen. Everything became simpler and more visual.

9.3.1. *The method for project creation*

The same operations are always used to develop a project using a manager and can be carried out in different orders depending on the software characteristics or the project creation method chosen by those in charge:

– defining general elements: start date, end date (retro-planning), units, general project calendars, various parameters for display, calculation and formats;

– potential break down into subprojects;

– entering tasks or activities and their duration;

– possible creation of a subgroup of tasks;

– entering antecedents/follow-ups;

– entering resource calendars;

– entering or sharing existing resources (name, group, type, costs, etc.);

– allocating resources to tasks;

– creating different dashboards;

– calculating and/or displaying margins;

– calculating and/or displaying charges;

---

1 This information is taken from the article focusing on Microsoft Project on Wikipedia: http://en.wikipedia.org/wiki/Microsoft_Project.

– possible manual or automatic leveling;

– potential online publication;

– implementing potential management of user rights and access;

– launching, monitoring, modifying and editing;

– creating, configuring and generating potential reports;

– possible data export to other software.

In the second part of this work (see Chapter 12), the suggested exercises will be carried out using Microsoft Project 2013 and can easily be adapted to other software.

Although this list is far from being exhaustive, we can site:

In Microsoft Windows:

– Sciforma PSN;

– Open workbench (open source);

– Gantt Project (free software);

– OpenProj (open source);

– Clarizen;

– Wrike (on line project management);

– Primavera Project management.

In Mac OSX:

– iTaskX;

– Merlin;

– FastTrack schedule;

– Xplan;

– ProjectLibre (GPL licence);

– Gantt Project (free software);

– OpenProj (open source).

In Android:

– Ganttdroid Lite (free);

– Ganttdroid Pro;

– GanttMan (free);

– GanttMan Pro;

– Project Schedule Free (free);

– Project Schedule.

In iOS:

– Project Planning Pro (free);

– Project Planner;

– Merlin – Project Management (free);

– Project Explore.

Online, via an Internet browser:

– Wrike;

– Clarizen – Online Project Management;

– Projectpro;

– Tom's Planner;

– Bitrix 24 (free for up to 12 users).

9.4. Flow simulators

These serve to enable the user to model a real system in order to see how a process or an activity will develop in different conditions. This technique enables different hypotheses or configurations to be modeled at low cost.

The model should be able to take account of all the existing constraints and launching the simulation shows the system's dynamic behavior while providing data that can be interpreted in the forms of graphs or dashboards or analyzed using a spreadsheet.

The user will thus be able to evaluate a real system's performance before it is developed or while it is performing to bring about potential improvements or modifications.

This type of software involves repetitive steps: modeling – experimenting/testing – interpreting results – improving/optimizing – testing – etc.

Modeling can take on many aspects, from a simple organization chart to two-dimensional (2D), or even three-dimensional (3D) animation.

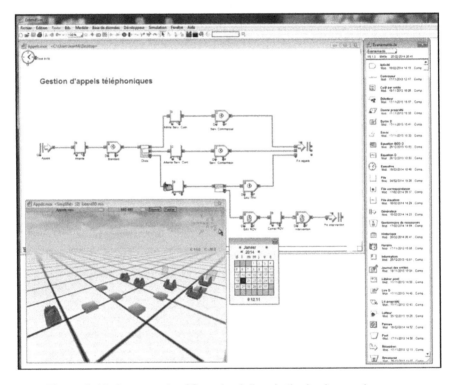

Figure 9.12. *An example of flow simulation. In the background, you can see the window that models the organization chart for the process and in the foreground, you can see the window for visualizing 3D simulation*

The software available commercially can be classified into two categories; general software that can handle all types of flux (people, materials, merchandise, product ranges, containers, individual parts, subsets, documents, etc.) and specialized software, adapted to specific flows and applications such as managing pedestrians, public transport, evacuating buildings, etc.

We can cite, and the list is not exhaustive: FlexSim, Simio, Simile, ExtendSim, Witness, Arenaand Simwalk.

In this work, the simulation exercises will be modeled with ExtendSim Suite version 9.

9.4.1. *Creating a simulation process*

Although simulation software is very different, the same stages, which we will explain below, are used to build simulation models:

– defining the main simulation parameters (units, the simulation's duration, speed, type, calendars, etc.);

– drawing the process model to be simulated. There could be several of these models if the process is broken down into subprocesses. In this case, a master model will group all the elements together;

– entering the constraints in the primitives or modeling blocks (times, durations, variables, etc.);

– defining in-going and out-going flows;

– creating resources and allocations;

– launching the simulation;

– creating personalized reports;

– development, enhancement and improvements;

– possible prioritization;

– possible presentation (integrating the model into a 2D or 3D graph).

Chapter 13 of this work focuses on computerized flow simulation using an example via ExtendSim.

10

Operational Research Using a Spreadsheet

10.1. Note

In this chapter, we will present several possible ways of using the Microsoft Excel spreadsheet to resolve classic problems associated with operational research.

The version that we use is part of the Microsoft Office 2013 suite. Most of the exercises shown will be easily adaptable to other spreadsheets provided that they have similar functions.

10.2. Dynamic programming

Here, we are going to tackle a problem encountered in section 4.2.3, the famous *knapsack problem* (KP).

We present a solution here based on an accounting spreadsheet associated with two procedures written in Visual Basic for Applications (VBAs).

To begin, you should create the table shown in Figure 10.1 in a spreadsheet.

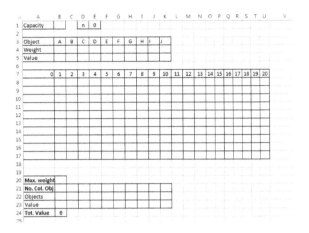

Figure 10.1. *The table to be created in a spreadsheet*

Once the data have been entered, you should place the following formulae in the cells indicated:

– E1: =COUNTA(B4:K4), calculates the number of objects available;

– B24: =SUM(B23:T23), calculates the objects' total value;

– then save your table;

– go into VBE, create a module for your project (DEVELOPER tab, VISUAL BASIC icon, right click on the project file, INSERT then MODULE);

– open the module that has just been created, usually MODULE 1, and enter the following lines of code:

```
Sub Clear() ↵
    'Clears the contents of the initial table ↵
rep = MsgBox("Do you wish to clear the tables' contents
to begin a new calculation?", vbYesNo, "Clear") ↵
    If rep = vbYes Then ↵
Range("B1").ClearContents ↵
Range("B3:K5").ClearContents ↵
Range("A8:U17").ClearContents ↵
Range("B20:K23").ClearContents ↵
    End If ↵
End Sub ↵
↵
```

```
Sub Knapsack() ↵
rep = MsgBox("Have you entered the capacity and values in
the table above to begin a new calculation?", vbYesNo,
"Calcul") ↵
    If rep = vbNo Then Exit Sub ↵
'Declaration of variables and tables
    Dim i, j, m1, m2 As Integer ↵
    'Table of weights ↵
    Dim p() As Integer ↵
    'Table of values ↵
    Dim v() As Integer ↵
    Dim c As Integer ↵
    'Allocation of the capacity to the variable c ↵
    c = Cells(1, 2) ↵
    Dim n As Integer ↵
    'Allocation of the number of objects to the variable
n ↵
    n = Cells(1, 5) ↵
ReDimp(n) ↵
ReDimv(n) ↵
    'Objects matrix ↵
    Dim table() As Integer ↵
ReDimtable(n, c) ↵
 ↵
    'Reading object weight ↵
    For i = 1 To n ↵
p(i) = Cells(4, i + 1) ↵
    Next ↵
↵
    'Reading object weight ↵
    For j = 1 To n ↵
v(j) = Cells(5, j + 1) ↵
    Next ↵
↵
    'Initialising the table, setting its content to 0 ↵
    For j = 0 To c ↵
table(0, j) = 0 ↵
    Next ↵
↵
'Constructing and filling the matrix ↵
    For i = 1 To n ↵
        For j = 0 To c ↵
            If j >= p(i) Then ↵
                m1 = table(i - 1, j) ↵
m2 = table(i - 1, j - p(i)) + v(i) ↵
If m1 > m2 Then table(i, j) = m1 Else table(i, j) = m2 ↵
            Else ↵
```

```
table(i, j) = table(i - 1, j)
            End If
Cells(7 + i, j + 1) = table(i, j)
        Next
    Next

'Reading objects
    Dim ob(8) As String
For u = 1 To n
ob(u) = Cells(3, u + 1)
Next

'Further reallocation of variables
    i = n
    j = c

    'Determining maximal weight
    Dim k, m As Integer
    m = 2
    For k = c To 0 Step -1
        If table(i, j) = table(i, j - 1) Then j = j - 1
    Next
Cells(20, 2) = j

    'Obtaining the list of objects in the table of
results
    While i > 0 And j > 0
        If table(i, j) <> table(i - 1, j) Then
j = j - p(i)
            Cells(22, m) = ob(i)
Cells(23, m) = v(i)
Cells(21, m) = i
i = i - 1
            m = m + 1
        Else
i = i - 1
End If
    Wend
End Sub
```

Figure 10.2. *The VBA code window in VBE attached to project Module 1. For a color version of the figure, see www.iste.co.uk/reveillac/logistics.zip*

To launch these two procedures, `Clear()` and `Knapsack()`, create two buttons by inserting two forms to which you assign each of the two macros (you should right click on forms, then ASSIGN MACRO⋯). Your spreadsheet should resemble the one in Figure 10.3.

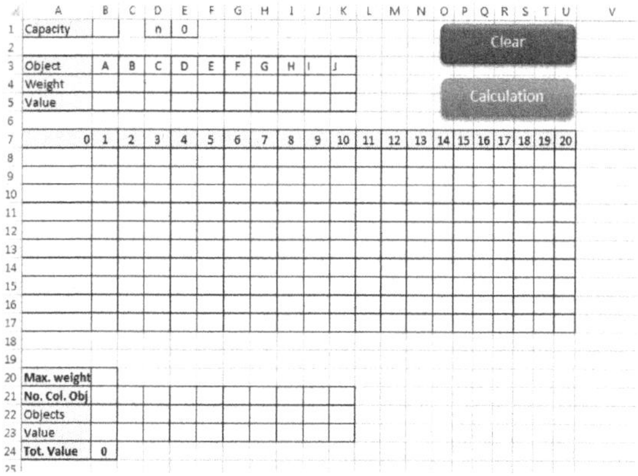

Figure 10.3. *The accounting spreadsheet containing two (Clear and Calculation) procedure buttons (macros)*

To use the KP calculation, you simply need to click on the CLEAR button to remove any calculations that have already been carried out and erase the calculated contents.

You should enter the data into the table above a second time (B1 and B3 to K5).

And finally, in the last phase, you should click on the CALCULATE button to launch the calculation and display the results.

	A	B	C	D	E	F	G	H	I	J	K	L	M	N	O	P	Q	R	S	T	U	V
1	Capacity	14		n	6											Clear						
2																						
3	Object	A	B	C	D	E	F	G	H	I	J											
4	Weight	2	1	5	2	4	3									Calculation						
5	Value	7	8	14	5	10	15															
6																						
7		0	1	2	3	4	5	6	7	8	9	10	11	12	13	14	15	16	17	18	19	20
8	0	0	7	7	7	7	7	7	7	7	7	7	7	7	7							
9	0	8	8	15	15	15	15	15	15	15	15	15	15	15								
10	0	8	8	15	15	15	22	22	29	29	29	29	29	29	29							
11	0	8	8	15	15	20	22	22	29	29	34	34	34	34								
12	0	8	8	15	15	20	22	25	29	30	34	34	39	44								
13	0	8	8	15	23	23	30	30	35	37	40	44	45	49	49							
14																						
15																						
16																						
17																						
18																						
19																						
20	Max. weight	13																				
21	No. Col. Obj	6	4	3	2	1																
22	Objects	F	D	C	B	A																
23	Value	15	5	14	8	7																
24	Tot. Value	49																				

Figure 10.4. *The example from section 4.2.3.3 calculated. You can see the results in the table below*

NOTE.– In the sheet that we have just created, we can handle a maximum of 10 objects. Nothing prevents you from increasing this number by adding the necessary modifications to the relevant VBA tables and procedures.

10.3. Scheduling

In this section, we are going to describe two possible uses for the Excel table:

– a critical path and project duration calculation method based on a calculation using a matrix (a double entry table);

– generating a simple Gantt chart from a precedence table.

10.3.1. *Critical path calculation matrix*

Using the intermediary of a Microsoft Excel spreadsheet, containing a matrix constructed around several calculation formulae, we will be able to determine a project's critical path and duration (see section 5.9).

The table shown is limited to a program evaluation and review technique (PERT) chart carrying a maximum of 10 points. It can easily be adapted to a higher number of lines and columns by applying the same logic to create the missing calculation formulae.

To demonstrate and explain how this spreadsheet functions, we will return to exercise 2 from Chapter 5 (section 5.15.2) the matrix of which is below.

	1	2	3	4	5	6	7	8	i
1		2							0
2			10	6	8				2
3					0				12
4					4	10	18		8
5							5		12
6							4		18
7								4	26
8									30
j	0	2	21	8	21	22	26	30	

Table 10.1. *The double entry matrix from exercise 2*

Now, let us create an accounting spreadsheet like that in Figure 10.5.

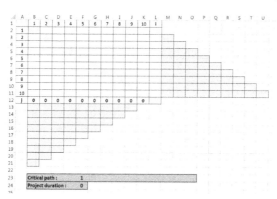

Figure 10.5. *The table to be created in the spreadsheet*

We can see that the matrix created is laid out for 10 points. The gray zones will receive the calculation formulae as well as the columns for the earliest start and latest end dates i and j.

Let us begin with the right-hand side, in Figure 10.6, you can see the formulae to be entered (only cells L3 to P6 are shown, the others follow the same logic).

J	K	L	M	N	O	P
9	10	i				
		0				
		=IF(COUNTA(D2:D3)=0,L2,M3)	=I2+C2			
		=IF(MAX(M4:S4)=0,L3,MAX(M4:S4))	=IF(D3<>"",L3+D3,"")	=IF(D2<>"",L2+D2,"")		
		=IF(MAX(M5:S5)=0,L4,MAX(M5:S5))	=IF(E4<>"",L4+E4,"")	=IF(E3<>"",L3+E3,"")	=IF(E2<>"",L2+E2,"")	
		=IF(MAX(M6:S6)=0,L5,MAX(M6:S6))	=IF(F5<>"",L5+F5,"")	=IF(F4<>"",L4+F4,"")	=IF(F3<>"",L3+F3,"")	=IF(F2<>"",L2+F2,"")

Figure 10.6. *The right-hand side with the calculation formulae to be entered*

We will comment on those formulae that fit the calculation technique described in section 5.9.3. They simply automate the step:

– M3: =L2+C2, adds the earliest start date for the first task to the first value encountered;

– L3: =IF(COUNTA(D2:D3)=0,L2,M3), checks for the absence of a value in D2 and D3, if the value is absent, it displays the previous earliest start date, which is L2, otherwise nothing is displayed;

– M4: =IF(D3<>"",L3+D3, ""), checks for the presence of a value in D3 and in this case effects the sum L3+D3, otherwise nothing is displayed;

– N4: =IF(D2<>"",L2+D2, ""), identical to the previous test for cell D2. The other group of gray cells, from M5 to U11, behaves in the same way, by taking account of each of the values present or absent in the column in question;

– L4: =IF(MAX(M4:S4)=0,L3,MAX(M4,S4)), checks the maximal value calculated in the column (the result of the calculations placed in cells M4 to S4). If it is equal to 0, no values will have been calculated, we can, therefore, assume that all the points have been reached, in which case the last value calculated from the beginning to the end should be copied, which is L3, otherwise the maximum is displayed. The same goes for the other formulae placed from L5 to L11 with the corresponding cells.

Let us now turn to the lower part, cells K12 to B21. As before, only some formulae are shown, since the rest follow the same logic:

– K12: =L11, returns to the last value in the column of earliest start dates;

– C13: =IF(D3<>"",D12-D3, ""), checks if the cell located below the diagonal is empty and if it is not, calculates the difference with the corresponding cell, which is D12-D3, otherwise nothing is displayed. This calculation is identical for all the other cells in the gray zone if the cell references are changed to obtain the appropriate result (see Figure 10.7);

– C12: =IF(MIN(C13:C21)=0,D12,MIN(C13:C21)), checks the minimal value calculated in the column (the result of the calculations in cells C13 to C20). If it is equal to 0, no values will have been calculated, we can, therefore, assume that the vertex has not been reached, in which case the last value calculated for the latest end date is copied, which is cell D12, otherwise the minimum is displayed. The same goes for the other formulae located in D12 to J12 with the corresponding cells;

– B12: =MIN(B13:B21), calculates the minimal value from zone B13 to B21, to display the latest end date of the first vertex in the graph.

	A	B	C	D
10	9			
11	10			
12	j	=MIN(B13:B21)	=IF(MIN(C13:C21)=0,D12,MIN(C13:C21))	=IF(MIN(D13:D21)=0,E12,MIN(D13:D21))
13		=IF(C2<>"",C12-C2,"")	=IF(D3<>"",D12-D3,"")	=IF(E4<>"",E12-E4,"")
14		=IF(D2<>"",D12-D2,"")	=IF(E3<>"",E12-E3,"")	=IF(F4<>"",F12-F4,"")
15		=IF(E2<>"",E12-E2,"")	=IF(F3<>"",F12-F3,"")	=IF(G4<>"",G12-G4,"")
16		=IF(F2<>"",F12-F2,"")	=IF(G3<>"",G12-G3,"")	=IF(H4<>"",H12-H4,"")

	I	J	K
	=IF(MIN(I13:I14)=0,J12,MIN(I13:I14))	=IF(MIN(J13)=0,K12,J13)	=L11
	=IF(J9<>"",J12-J9,"")	=IF(K10<>"",K12-K10,"")	
	=IF(K9<>"",K12-K9,"")		

Figure 10.7. *The cells below the CPM matrix with their formulae*

Let us now focus on finding the critical path through the graph node numbers. The project duration is associated with the last value for the earliest beginning, which is L11:

– F23: =IF(B12=L2,B1, ""), tests if cell B12 is equal to cell B2 and displays the node number located in B1. This formula is not compulsory, it can simply be replaced by =L2, since the first node is always in the critical path, except in specific cases;

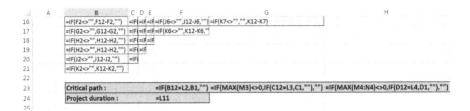

	A	B	C	D	E	F	G	H
16		=IF(F2<>"",F12-F2,"")	=IF	=IF	=IF(J6<>"",J12-J6,"")	=IF(K7<>"","",K12-K7)		
17		=IF(G2<>"",G12-G2,"")	=IF	=IF	=IF(K6<>"",K12-K6,"			
18		=IF(H2<>"",H12-H2,"")	=IF	=IF	=IF			
19		=IF(H2<>"",H12-H2,"")	=IF	=IF				
20		=IF(J2<>"",J12-J2,"")	=IF					
21		=IF(K2<>"",K12-K2,"")						
22								
23	Critical path :		=IF(B12=L2,B1,"")	=IF(MAX(M3)<>0,IF(C12=L3,C1,""),"")	=IF(MAX(M4:N4)<>0,IF(D12=L4,D1,""),"")			
24	Project duration :	=L11						
25								

Figure 10.8. *The calculation for the critical path and the duration displayed*

– G23: =IF(MAX(M3)<>0,IF(C12=L3,C1,"")‚""), we find
two interlinked tests. The first test verifies that M3 is different to 0, if this
was not the case, this would mean that the current node does not belong to
the critical path and nothing would be displayed. If M3 is different to 0,
then we check that C12 is equal to its corresponding cell, L3, on the
diagonal in the matrix and the node number C1 is displayed, otherwise
nothing is displayed;

– H23: =IF(MAX(M4:N4)<>0,IF(D12=L4;D1,"")‚""), identical
to the above, but with a zone of M4 to N4 to test the maximum;

– the same principle is then applied for the cells from I23 to O23 with
the corresponding cell references.

Here is the result obtained after entering the figures from Table 10.1.

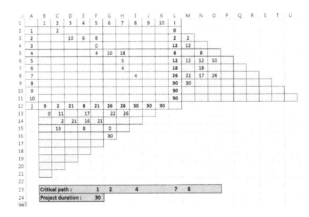

Figure 10.9. *The spreadsheet with the group of cells entered in the matrix. We can
see the values from i (earliest start) and from j (latest end) as well as the project
duration and the critical path, which pass through nodes: 1, 2, 4, and 7 and 8*

10.3.2. *Classic Gantt diagram*

To create this diagram, we will return to the public address system exercise discussed in section 5.6. In the precedence table, only the columns for tasks and duration (in minutes) are retained and we add two columns, beginning and end, to these.

We will enter this table into a Microsoft Excel spreadsheet, then we will add a variance column which calculates the difference between "End" and "Start" (in cell E2: =D2-B2) in order to obtain the matrix in Figure 10.10.

Task	Beginning	Duration	End
A	0	120	120
B	120	20	145
C	120	15	140
D	120	20	140
E	140	10	155
F	140	15	155
G	155	25	180
H	140	20	180
I	180	15	195

Table 10.2. *Tasks, beginnings, durations and ends for the sound system exercise*

	A	B	C	D	E	F
1	Task	Start	Duration	End	Variation	
2	A	0	120	120	120	
3	B	120	20	145	25	
4	C	120	15	140	20	
5	D	120	20	140	20	
6	E	140	10	155	15	
7	F	140	15	155	15	
8	G	155	25	180	25	
9	H	140	20	180	40	
10	I	180	15	195	15	
11						
12						

Figure 10.10. *The matrix corresponding to the table of tasks with all its columns*

Then, select the table, and then carry out the following group of operations:

– INSERT tab, CHARTS toolbar, scroll down to INSERT BAR CHART, choose STACKED BAR in 2-D BAR;

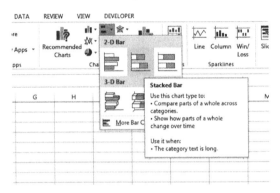

Figure 10.11. *Microsoft Excel: to insert a 2D STACKED BAR CHART*

– then move and resize the graph;

Figure 10.12. *The resized graph placed beneath the matrix. For a color version of the figure, see www.iste.co.uk/reveillac/logistics.zip*

– click on the y-axis;

– right click, choose FORMAT AXIS;

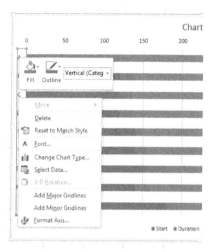

Figure 10.13. *The pop-up menu (right click) to format the axis. For a color version of the figure, see www.iste.co.uk/reveillac/logistics.zip*

– in the window FORMAT AXIS, check CATEGORIES IN REVERSE ORDER;

Figure 10.14. *The axis configuration window in Microsoft Excel 2013, with its checkbox: values in reverse order (under Axis Options)*

– click on the x-axis to select it;

– in the window FORMAT AXIS, open LABELS and chose HIGH in LABEL POSITION;

– enter 10 and 5, respectively, in AXIS OPTIONS as MAJOR and MINOR UNITS;

Figure 10.15. *The axis configuration (10 and 5) and labels (above)*

– right click on the bottom of the graph and choose SELECT DATA;

– in the data source window that opens, uncheck "Duration" and "End Date" (you can also make these disappear via the DELETE button), then confirm by clicking on the OK button;

Figure 10.16. *The unchecked "Duration" and "End" fields in the window displaying the data source. For a color version of the figure, see www.iste.co.uk/reveillac/logistics.zip*

– click on the legend and delete it (DELETE key or backspace on Mac);

– right click on the "start" bars in the graph, select NO OUTLINE and NO FILL by clicking on the appropriate icons (LINE and FILL) in the pop-up tool bar;

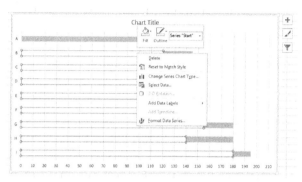

Figure 10.17. *The menu and the pop-up tool bar, obtained by right clicking on the "Start" bar. You can see the LINE and FILL tools above the menu*

– double-click on the title to change it by typing "Outdoor public address system";

– to exit, save your work.

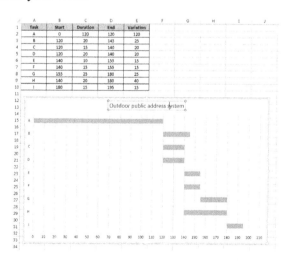

Figure 10.18. *The completed Gantt chart with its corresponding table in a Microsoft Excel 2013 spreadsheet*

10.3.3. *A Gantt chart with a calendar*

When dates need to be managed according to a calendar that can show working days and holidays, we should consider excluding the latter from the project timeline.

Below, you will find an example that relies on these conditions.

Installing an air-conditioning system (simplified plan):

Task	Designation	Duration (days)	Precedence
A	Survey of requirements (electrical equipment, piping, fittings, refrigeration gas and various other components)	1	–
B	Order and delivery of piping, fittings and various components	10	A
C	Order and delivery of electrical equipment	5	A
D	Order and delivery of diffusers and group exchanger	8	A
E	Order and delivery of refrigerating gas	4	A
F	Assembly, installation and laying of the group exchanger	2	D
G	Installation and laying diffusers	4	D
H	Laying and connecting pipes	2	B, F, G
I	Electrical cable and connection	1	C, F, G
J	Filling, pressurizing and purging the installation	1	E, H
K	Starting and testing the installation and adjustments	1	H, I, J
L	Clearing the work site	1	K

Table 10.3. *The precedence table for installing an air-conditioning system*

For this second exercise, we will assume that the project start date is Monday March 2nd 2015.

First, a table like that shown in Figure 10.19 should be entered in Microsoft Excel. The start and end dates have been calculated in the standard way using a PERT or MPM method (the MPM is shown in Figure 10.20).

	A	B	C	D	E	F
1	Project start date :		03/02/15			
2						
3	Task	Start	Duration	End	Variation	
4	A	03/02/15	1	03/03/15	1	
5	B	03/03/15	10	03/17/15	14	
6	C	03/03/15	5	03/10/15	7	
7	D	03/03/15	8	03/13/15	10	
8	E	03/03/15	4	03/09/15	6	
9	F	03/11/15	2	03/13/15	2	
10	G	03/11/15	4	03/17/15	6	
11	H	03/15/15	2	03/17/15	2	
12	I	03/15/15	1	03/16/15	1	
13	J	03/17/15	1	03/18/15	1	
14	K	03/18/15	1	03/19/15	1	
15	L	03/19/15	1	03/20/15	1	

Figure 10.19. *The table to be created in Microsoft Excel (here the critical tasks are in bold, shaded type)*

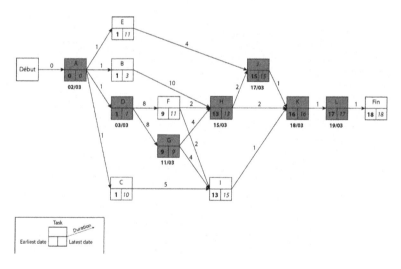

Figure 10.20. *The MPM diagram for the project: installing an air-conditioning system*

In cell D4 (End), we have the formula =B4+C4, which we fill downward as far as D15.

In cell E4 (Variance), we have the formula =D4-B4, which we fill downward as far as E15.

We can see that the end dates do not take account of 5-day weeks where only the days from Monday to Friday are working days. To overcome this problem, we can replace the formula located in D4 with =SERIES.WORK.DAY(B4;C4). We then obtain the table in Figure 10.21.

	A	B	C	D	E	F
1	Project start date :		03/02/15			
2						
3	Task	Start	Duration	End	Variation	
4	A	03/02/15	1	03/03/15	1	
5	B	03/03/15	10	03/17/15	14	
6	C	03/03/15	5	03/10/15	7	
7	D	03/03/15	8	03/13/15	10	
8	E	03/03/15	4	03/09/15	6	
9	F	03/11/15	2	03/13/15	2	
10	G	03/11/15	4	03/17/15	6	
11	H	03/15/15	2	03/17/15	2	
12	I	03/15/15	1	03/16/15	1	
13	J	03/17/15	1	03/18/15	1	
14	K	03/18/15	1	03/19/15	1	
15	L	03/19/15	1	03/20/15	1	

Figure 10.21. *The table recalculated for working days*

To display the graph correctly, it will be necessary to change the format of the dates in the "Start" and "End" columns by replacing the date format with the standard format:

– select cells B4 to B15, then, while holding down CTRL, cells D4 to D15;

– right click, FORMAT CELLS…, NUMBER tab, GENERAL category.

We then obtain the number of days passed since 1st January 1900 in each of the cells.

	A	B	C	D	E	F
1	Project start date :		03/02/15			
2						
3	Task	Start	Duration	End	Variation	
4	A	42065	1	42066	1	
5	B	42066	10	42080	14	
6	C	42066	5	42073	7	
7	D	42066	8	42076	10	
8	E	42066	4	42072	6	
9	F	42074	2	42076	2	
10	G	42074	4	42080	6	
11	H	42078	2	42080	2	
12	I	42078	1	42079	1	
13	J	42080	1	42081	1	
14	K	42081	1	42082	1	
15	L	42082	1	42083	1	

Figure 10.22. *The "Start" and "End" columns in STANDARD format*

Then, select the table, and then carry out the following group of operations:

– INSERT tab, CHARTS tool bar, scroll down to BAR, choose STACKED BAR under 2-D BAR;

– then move and resize the graph;

– click on the y-axis;

– right click, choose FORMAT AXIS···;

– in the FORMAT AXIS window, check CATEGORIES IN REVERSE ORDER;

– click on the x-axis to select it;

– in the FORMAT AXIS window, click on the histogram-shaped icon;

– enter 42065 (the value of cell B4, which is 03/02/2015, the project start date) in AXIS OPTIONS as the MINIMUM;

– then, enter 7 and 1, respectively, as MAJOR and MINOR UNITS;

– open LABELS and choose HIGH in LABEL POSITION;

– open NUMBER, choose DATE in CATEGORY then *3/14//2001 in TYPE;

Figure 10.23. *AXIS OPTIONS, LABELS and NUMBER*

– in the FORMAT AXIS window, click on the cross-shaped icon (to the left of the histogram);

– open ALIGNMENT and enter –45° in the field CUSTOM ANGLE in order to rotate the dates on the x-axis;

Figure 10.24. *AXIS OPTIONS, ALIGNMENT at –45°*

– right click on the bottom of the chart and choose SELECT DATA;

– in the "Data Sources" window that opens, uncheck "Duration" and "End" (you can also make these disappear via the DELETE button) and then confirm by clicking the OK button;

– click on the legend and delete it (press DEL key or backspace on a Mac);

– right click on the "Start" bars in the graph, choose NO LINE and NO FILL by clicking on the relevant icons (LINE and FILL) in the pop-up tool bar;

– right click on a date on the x-axis, ADD A MINOR GRIDLINE;

– double-click on the title to change it by typing "Installing an air conditioning system";

– then, reapply a date format in the "Start" and "End" columns;

– to exit, save your work;

– then, you can change the colors in the chart, for example, to show the critical tasks in a different color.

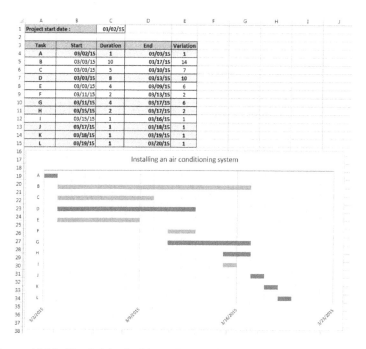

Figure 10.25. *The finished table and graph for the project Gantt chart. For a color version of the figure, see www.iste.co.uk/reveillac/logistics.zip*

The exercises on creating a Gantt chart studied previously can be adapted to more sizeable projects. The table size will be larger but all the principles and techniques to be used remain the same.

10.4. Maximal flows

To calculate the maximal flow available in a network between the source and sink, we will use the Excel solver along with VBA procedures and a spreadsheet. In the solution shown, the number of edges (or vertices) is limited to 25 and the number of vertices is limited to 20.

Begin by creating the spreadsheet in Figure 10.26.

Figure 10.26. *The spreadsheet to be created in Microsoft Excel*

Then, enter the following calculation formulae:

– G2: =IF(F2="","",SUMIF(From,F2,Flow)-SUMIF(To,F2, Flow)), if the cell in the "Nodes" column is empty nothing is displayed, otherwise the difference is calculated: the sum of flows equal to F2 in the "From" column – the sum of flows equal to F2 in the "To" column. This determines the net flow;

– "From" and "To" are cell ranges that will be named later by a VBA procedure;

– copy G2 downward as far as cell G21;

– C27: =COUNTA(A2:A26), counts the number of edges;

– G22: =COUNTA(F2:F21), counts the number of vertices;

– C28: =G2, displays the maximal flow;

– then, save your table;

– go into VBE, create a module for your project (DEVELOPER tab, VISUAL BASIC icon, right click on the project file, INSERT then MODULE);

– open the module you have just created, usually MODULE 1 and enter the following lines of code:

```
Sub Design_tables() ↵
Dim CountArc, CountNodesAs String ↵
'Clear tables ↵
'If  click on No then clear otherwise quit ↵
    If MsgBox("Do you wish to clear the table content?",
vbYesNo, "Clear") = vbYes Then ↵
'Clears the cell content ↵
Range("A2:D26,F2:F21,H3:H21").Select ↵
Selection.ClearContents ↵
        'Deletes the cell background color ↵
Range("A2:D26,F2:G21,H3:H21").Select ↵
Selection.Interior.ColorIndex = 0 ↵
Range("C28").Select ↵
        'Relaunch automatic sheet recalculation ↵
Application.Calculation = xlCalculationAutomatic ↵
Else ↵
        Exit Sub ↵
End If ↵
'Name and color allocation ↵
CountArc = InputBox("How many edges does your graph have?",
"Number of edges") ↵
    'If click on Cancel then quit ↵
    If CountArc = "" Then Exit Sub ↵
Range(Cells(2, 1), Cells(1 + CountArc, 1)).Name = "From" ↵
Range(Cells(2, 1), Cells(1 + CountArc,
3)).Interior.ColorIndex = 40 ↵
Range(Cells(2, 2), Cells(1 + CountArc, 2)).Name = "To" ↵
Range(Cells(2, 2), Cells(1 + CountArc,
3)).Interior.ColorIndex = 44 ↵
Range(Cells(2, 3), Cells(1 + CountArc, 3)).Name = "Flow" ↵
Range(Cells(2, 3), Cells(1 + CountArc,
3)).Interior.ColorIndex = 36 ↵
Range(Cells(2, 4), Cells(1 + CountArc, 4)).Name = "Capacity" ↵
Range(Cells(2, 4), Cells(1 + CountArc,
4)).Interior.ColorIndex = 27 ↵
CountNode = InputBox("How many vertices does your graph
have?", "Number of vertices") ↵
    'If click on Cancel then quit ↵
    If CountNode = "" Then Exit Sub ↵
Range(Cells(2, 6), Cells(1 + CountNode, 6)).Name = "Node" ↵
```

```
Range(Cells(2, 6), Cells(1 + CountNode,
6)).Interior.ColorIndex = 17
Range(Cells(3, 7), Cells(CountNode, 7)).Name = "NetFlow"
Range(Cells(2, 7), Cells(1 + CountNode,
7)).Interior.ColorIndex = 37
Range(Cells(3, 8), Cells(CountNode, 8)).Name =
"SupplyDemand"
Range(Cells(3, 8), Cells(CountNode, 8)).Interior.ColorIndex
= 8
MsgBox ("You can now enter your" &CountArc& " arcs with
their capacity in the FROM and TO column. Then enter the
list of your " &CountNode& "nodes in VERTICES column.")
Range(Cells(3, 8), Cells(CountNode, 8)) = 0
End Sub

'Launch the solver to find the solution
Sub Resolve()
    'Reset solver to zero
SolverReset
    'Defining solver parameters
SolverOkSetCell:="$C$28", MaxMinVal:=1, ValueOf:=0,
ByChange:="$C$2:$C$19", _
        Engine:=2, EngineDesc:="Simplex LP"
SolverDeleteCellRef:="$C$2:$C$19", Relation:=1,
FormulaText:="Capacity"
SolverOkSetCell:="$C$28", MaxMinVal:=1, ValueOf:=0,
ByChange:="$C$2:$C$19", _
        Engine:=2, EngineDesc:="Simplex LP"
SolverAddCellRef:="$C$2:$C$19", Relation:=1,
FormulaText:="Capacity"
SolverOkSetCell:="$C$28", MaxMinVal:=1, ValueOf:=0,
ByChange:="$C$2:$C$19", _
        Engine:=2, EngineDesc:="Simplex LP"
SolverDeleteCellRef:="$G$3:$G$10", Relation:=2,
FormulaText:="SupplyDemand"
SolverOkSetCell:="$C$28", MaxMinVal:=1, ValueOf:=0,
ByChange:="$C$2:$C$19", _
        Engine:=2, EngineDesc:="Simplex LP"
SolverAddCellRef:="$G$3:$G$10", Relation:=2,
FormulaText:="SupplyDemand"
SolverOkSetCell:="$C$28", MaxMinVal:=1, ValueOf:=0,
ByChange:="$C$2:$C$19", _
        Engine:=2, EngineDesc:="Simplex LP"
SolverOkSetCell:="$C$28", MaxMinVal:=1, ValueOf:=0,
ByChange:="$C$2:$C$19", _
        Engine:=2, EngineDesc:="Simplex LP"
    'Launch automatic calculation of results by the solver
SolverSolve (True)
End Sub
```

To launch these two procedures, Design_tables()and Solve(), create two buttons by inserting two forms to which you assign each of the two macros (right click on form, then ASSIGN MACRO···). Your spreadsheet should resemble the one in Figure 10.27.

Figure 10.27. *The sheet, with two buttons ("Design tables" and "Solve")*

To use the maximal flow calculator, it should be enough to click on the button CREATE TABLES to remove any calculations already carried out and clear the calculated contents. Two dialogue boxes will appear one after the other, asking the number of edges and then the number of vertices on your graph.

Enter the data in the table on the left including, for each edge, the starting vertex in the "From" column and its end vertex in the "To" column followed by its capacity in the "Capacity" column.

Then, add the flows; we suggest you to add the minimal value for capacity for all the flows.

Then, enter all the vertices in the table on the right, in the "Vertices" column beginning with the source (usually S) and ending with the sink (usually T), the order is not important for the other vertices.

You can see that the entry fields have been colored to make data entry easier.

To finish, click the SOLVE button, to launch solution calculation by the solver and display the results.

In Figure 10.28, we return to the data from exercise 1, Chapter 6 (section 6.5.1), you can see all the edges and their capacities (18), all the vertices (10) and the starting flow, which dates from 2000.

	A	B	C	D	E	F	G	H
1	**From**	**To**	**Flow**	**Capacity**		**Nodes**	**Net Flow**	**Supply/Demand**
2	S	P1	2000	10000		S	6000	
3	S	P2	2000	8000		P2	0	0
4	S	P3	2000	6000		P3	2000	0
5	P1	P2	2000	4000		R1	0	0
6	P1	R1	2000	6000		R2	-2000	0
7	P2	R1	2000	4000		R3	0	0
8	P2	R2	2000	2000		R4	-2000	0
9	P2	R3	2000	2000		R5	-2000	0
10	P3	P2	2000	4000		P1	2000	0
11	P3	R3	2000	6000		T	-4000	
12	R1	R2	2000	4000				
13	R1	R4	2000	6000				
14	R2	R4	2000	4000				
15	R2	R5	2000	3000				
16	R3	R2	2000	6000				
17	R3	R5	2000	2000				
18	R4	T	2000	12000				
19	R5	T	2000	4000				
20								
21								
22						No. Nodes	10	
23								
24								
25								
26								
27	No. Edges		18			Create tables		
28	**Maximum flow**		**6000**					Solve
29								

Figure 10.28. *The data from exercise 1, Chapter 6, entered on a spreadsheet. For a color version of the figure, see www.iste.co.uk/reveillac/logistics.zip*

In Figure 10.29, you can see the results of the calculation once it has been launched by clicking on the SOLVE button.

	A	B	C	D	E	F	G	H
1	From	To	Flow	Capacity		Nodes	Net Flow	Supply/Demand
2	S	P1	6000	10000		S	14000	
3	S	P2	8000	8000		P2	0	0
4	S	P3	0	6000		P3	0	0
5	P1	P2	0	4000		R1	0	0
6	P1	R1	6000	6000		R2	0	0
7	P2	R1	4000	4000		R3	0	0
8	P2	R2	2000	2000		R4	0	0
9	P2	R3	2000	2000		R5	0	0
10	P3	P2	0	4000		P1	0	0
11	P3	R3	0	6000		T	-14000	
12	R1	R2	4000	4000				
13	R1	R4	6000	6000				
14	R2	R4	4000	4000				
15	R2	R5	2000	3000				
16	R3	R2	0	6000				
17	R3	R5	2000	2000				
18	R4	T	10000	12000				
19	R5	T	4000	4000				
20								
21								
22						No. Nodes	10	
23								
24								
25								
26								
27	No. Edges		18			Create tables		Solve
28	Maximum flow		14000					
29								

Figure 10.29. *The solution for exercise 1 with all the results displayed. For a color version of the figure, see www.iste.co.uk/reveillac/logistics.zip*

10.5. Transport model

Transport problems are common in logistics, it is often necessary to harmonize the distances, costs and constraints of supplying customers.

Below, you will find two classic examples that will be solved by the Microsoft Excel Solver.

Using this as a basis, it will be easy for you to adapt each of the solutions to solve your own problems.

10.5.1. *Delivery to customers*

A freight train company must transport two batches of palletized products, departing from two departure stations (DS1 and DS2), to three destination stations (ES1, ES2 and ES3) distributed across France. The products will be transported via intermediate marshalling yards (MY1, MY2, MY3, MY4 and MY5).

The number of palettes available in the stations that the goods will depart from and the storage capacity in the end stations are known as well as the transport capacities of the track between the stations.

Table 10.4 groups the information for all the stations.

Palette flow										
→	DS1	DS2	MY1	MY2	MY3	MY4	MY5	ES1	ES2	ES3
Palettes available	105	75								
DS1		-	60	45	36	-	-	-	-	-
DS2	-		-	18	66	-	-	-	-	-
MY1	-	-	-	30	-	45	-	-	-	-
MY2	-	-	-	-	-	-	-	30	45	45
MY3	-	-	-	-	-		66	-	-	-
MY4	-	-	-	-	-	-	-	21	30	-
MY5	-	-	-	-	-		-	-	30	30
Storage capacity at end stations								45	45	60

Table 10.4. *Synthesis of palletized product batch flow*

If we look at the problem more closely, we can see that there is a problem involving maximal flow with constraints being exceeded.

We will tackle it using a simple double entry table, which involves using the solver.

First, reproduce and fill the table from Figure 10.30 by entering it into a spreadsheet.

All the listed data are shown. To this end, the table has an S column and a T column, which represent the source and sink, thus showing the number of palettes available at departure (DS1: 105 and DS: 75) and the end stations' storage capacity (ES1: 45, ES2: 45 and ES3: 60).

	A	B	C	D	E	F	G	H	I	J	K	L	M	N
1														
2														
3		DS : Departure station				MY : Marshalling yard				ES : End station			Flow :	0
4														
5		S	DS1	DS2	MY1	MY2	MY3	MY4	MY5	ES1	ES2	ES3	T	
6	S	0	105	75	0	0	0	0	0	0	0	0	0	
7	DS1	0	0	0	60	45	36	0	0	0	0	0	0	
8	DS2	0	0	0	0	18	66	0	0	0	0	0	0	
9	MY1	0	0	0	0	30	0	45	0	0	0	0	0	
10	MY2	0	0	0	0	0	0	0	0	30	45	45	0	
11	MY3	0	0	0	0	0	0	0	66	0	0	0	0	
12	MY4	0	0	0	0	0	0	0	0	21	30	0	0	
13	MY5	0	0	0	0	0	0	0	0	0	30	30	0	
14	ES1	0	0	0	0	0	0	0	0	0	0	0	45	
15	ES2	0	0	0	0	0	0	0	0	0	0	0	45	
16	ES3	0	0	0	0	0	0	0	0	0	0	0	60	
17	T	0	0	0	0	0	0	0	0	0	0	0	0	
18														

Figure 10.30. *The table to be constructed using a Microsoft Excel spreadsheet*

Under this first table, add a second table as shown in Figure 10.31.

		S	DS1	DS2	MY1	MY2	MY3	MY4	MY5	ES1	ES2	ES3	T	Total
20		S	DS1	DS2	MY1	MY2	MY3	MY4	MY5	ES1	ES2	ES3	T	Total
21	S	0	0	0	0	0	0	0	0	0	0	0	0	
22	DS1	0	0	0	0	0	0	0	0	0	0	0	0	0
23	DS2	0	0	0	0	0	0	0	0	0	0	0	0	0
24	MY1	0	0	0	0	0	0	0	0	0	0	0	0	0
25	MY2	0	0	0	0	0	0	0	0	0	0	0	0	0
26	MY3	0	0	0	0	0	0	0	0	0	0	0	0	0
27	MY4	0	0	0	0	0	0	0	0	0	0	0	0	0
28	MY5	0	0	0	0	0	0	0	0	0	0	0	0	0
29	ES1	0	0	0	0	0	0	0	0	0	0	0	0	0
30	ES2	0	0	0	0	0	0	0	0	0	0	0	0	0
31	ES3	0	0	0	0	0	0	0	0	0	0	0	0	0
32	T	0	0	0	0	0	0	0	0	0	0	0	0	
33	Total		0	0	0	0	0	0	0	0	0	0	0	
34														

Figure 10.31. *The second table, placed below the first*

It should be filled as shown below:

– line 33 contains the sum of each of the columns;

– in cell C33, enter: =SUM(C21:C32;

– then, copy this formula down to cell L33;

– column N contains the sum of each of the lines;

– in cell N22, enter: =SUM(B22:M22);

– then, copy this formula down to N31;

– in cell N3, place the sum of the sinks to be maximized, which is: =SUM(M21:M32);

– launch the solver, DATA tab, SOLVER tool in the tool ribbon. If it is not visible, consult Appendix 1 of this work to proceed with the installation;

– set the solver parameters as shown in Figure 10.32;

– SET OBJECTIVE: $N$3;

– BY CHANGING VARIABLE CELLS: $B$21:$M$32;

– SUBJECT TO THE CONSTRAINTS: $B$21:$M$32 <= $B$6:$M$17;

– SUBJECT TO THE CONSTRAINTS: $C$33:$L$33 = $N$22:$N$31;

– check MAKE UNCONSTRAINED VARIABLES NON-NEGATIVE;

– SELECT A SOLVING METHOD: Simplex PL;

Figure 10.32. *Setting the solver parameters*

– then, click on the SOLVE button, after a short calculation time, the solver will display the result window;

– you will then be able to SAVE SOLVER SOLUTION or even RESET INITIAL VALUES (to begin again) by checking the appropriate choice and clicking the OK button.

You will obtain the values from Figure 10.33 in the second table, which will specify the number of palettes to be transported between each point to fit all the constraints.

		S	DS1	DS2	MY1	MY2	MY3	MY4	MY5	ES1	ES2	ES3	T	Total
21	S	0	105	45	0	0	0	0	0	0	0	0	0	
22	DS1	0	0	0	57	45	3	0	0	0	0	0	0	105
23	DS2	0	0	0	0	18	27	0	0	0	0	0	0	45
24	MY1	0	0	0	0	30	0	27	0	0	0	0	0	57
25	MY2	0	0	0	0	0	0	0	0	30	33	30	0	93
26	MY3	0	0	0	0	0	0	0	30	0	0	0	0	30
27	MY4	0	0	0	0	0	0	0	0	15	12	0	0	27
28	MY5	0	0	0	0	0	0	0	0	0	0	30	0	30
29	ES1	0	0	0	0	0	0	0	0	0	0	0	45	45
30	ES2	0	0	0	0	0	0	0	0	0	0	0	45	45
31	ES3	0	0	0	0	0	0	0	0	0	0	0	60	60
32	T	0	0	0	0	0	0	0	0	0	0	0	0	
33	Total		105	45	57	93	30	27	30	45	45	60		

Figure 10.33. *The results calculated by the solver*

You can see that the total number of palettes arriving at the destination does not exceed 150 (45+45+60), the storage capacity at the end stations is, therefore, taken into account.

10.5.2. Transport at minimum cost

In this type of problem, we seek to convey goods at minimal cost while taking account of constraints such as customer demand and depot storage capacity.

Let us imagine three factories located (U1, U2 and U3) in France that want to provide for their clients (C1 to C5) distributed across Europe, the transport costs (per ton: 1,000 kg) are specified in Table 10.5.

Factory stock	C1	C2	C3	C4	C5
U1	$ 300.00	$2 400.00	$300.00	$1 500.00	$1 200.00
U2	$1 500.00	$1 500.00€	$900.00	$1 800.00	$2 100.00
U3	$ 600.00	$ 900.00	$1 500.00	$2 700.00	$2 400.00

Table 10.5. *Cost of transport between factories and clients*

Each factory has a volume of available products in stock, evaluated at: 7,200 kg for U1, 4,800 kg for U2 and 7,800 kg for U3.

The five clients would like the following quantities, C1: 3,600 kg, C2: 3,900 kg, C3: 4,350 kg, C4: 3,750 kg and C5: 4,200 kg.

We now have to define what product masses each of the three factories should dispatch to these five clients to meet their requests, taking account of the available stock and minimizing the delivery cost.

First, create the two tables from Figure 10.34 in a Microsoft Excel spreadsheet.

	A	B	C	D	E	F	G	H
1		Transport costs						
2		C1	C2	C3	C4	C5	Factory stock (kg)	
3	U1	$ 300.00	$2,400.00	$ 300.00	$1,500.00	$1,200.00	7200	
4	U2	$1,500.00	$1,500.00	$ 900.00	$1,800.00	$2,100.00	4800	
5	U3	$ 600.00	$ 900.00	$1,500.00	$2,700.00	$2,400.00	7800	
6	Order (Kg)	3600	3900	4350	3750	4200		
7								
8		C1	C2	C3	C4	C5	Quantity to be send	
9	U1						0	
10	U2						0	
11	U3						0	
12	Quantity to be received (kg)	0	0	0	0	0		
13								
14	Objective	0						
15								
16								
17								

Figure 10.34. *The two cost calculation tables*

Then, apply the following group of operations:

– in cell `B12`, enter the calculation: `=SUM(B9:B11)`;

– then, copy this formula to the right as far as `F12`;

– in G9, enter: =SUM(B9:F9);

– then, copy this sum downward, as far as G11;

– in cell B14, we will specify the calculation objective for the solver. This is the sum that we need to minimize;

– enter: =SUMPRODUCT(B3:F5,B9:F11);

– launch the solver, DATA tab, SOLVER tool in the ribbon. If it is not visible, consult Appendix 1 for the installation procedure;

– enter the parameters, as shown in Figure 10.35;

– SET OBJECTIVE: $B$14;

– BY CHANGING VARIABLE CELLS: $B$9:$F$11;

– SUBJECT TO THE CONSTRAINTS: $B$12:$F$12 = $B$6:$F$6;

– SUBJECT TO THE CONSTRAINTS: $G$9:$G$11 = $G$3:$G$5;

– check MAKE UNCONSTRAINED VARIABLES NON-NEGATIVE;

– SELECT A SOLVING METHOD: Simplex PL;

Figure 10.35. *Setting solver parameters*

– then, click on the SOLVE button and wait a few seconds;

– a window will open. You will then be able to SAVE THE SOLVER SOLUTION or even RESET ORIGINAL VALUES (to begin again) by checking the appropriate choice and clicking on the OK button.

The lower table displays the quantities to be dispatched from each of the factories to each of the clients. We can see that the constraints on the clients' orders and on the factory stocks have been met.

	A	B	C	D	E	F	G	H
1						Transport costs		
2		C1	C2	C3	C4	C5	Factory stock (kg)	
3	U1	$ 300.00	$2,400.00	$ 300.00	$1,500.00	$1,200.00	7200	
4	U2	$1,500.00	$1,500.00	$ 900.00	$1,800.00	$2,100.00	4800	
5	U3	$ 600.00	$ 900.00	$1,500.00	$2,700.00	$2,400.00	7800	
6	Order (Kg)	3600	3900	4350	3750	4200		
7								
8		C1	C2	C3	C4	C5	Quantity to be send	
9	U1	0	0	3300	0	3900	7200	
10	U2	0	0	1050	3750	0	4800	
11	U3	3600	3900	0	0	300	7800	
12	Quantity to be received (kg)	3600	3900	4350	3750	4200		
13								
14	Objective	19755000						
15								

Figure 10.36. *The results obtained after launching the solver*

10.6. Linear programming

We are now going to show you how to use the Microsoft Excel solver to solve classic linear programming problems, which we will resolve on paper with the help of the simplex.

We are now going to create a spreadsheet able to support as many as six constraints linked to six variables, while responding to maximization or minimization demands.

You can very easily increase the amount of data managed by applying the same calculation formulae on a larger table.

10.6.1. *Creating calculation tables*

In a spreadsheet, construct and fill the table from Figure 10.37, and then carry out the following steps:

	A	B	C	D	E	F	G	H	I	J	K
1	Variables	x1	x2	x3	x4	x5	x6				
2	bi										
3											
4	Constraints										
5	n1							0	<=		
6	n2							0	<=		
7	n3							0	<=		
8	n4							0	<=		
9	n5							0	<=		
10	n6							0	<=		
11											
12	z	20	30								
13											
14	Max z	0									
15											

Figure 10.37. *The table that will accommodate our linear programs*

– in cell H5, enter the formula: =SUMPRODUCT(B5:G5;$B$2:$G$2). This formula calculates the sum of the products between the cells in line 5 (constraint no. 1) and line 2 (b_i);

– copy this formula again downward, as far as H10;

– in cell B14, enter the formula =SUMPRODUCT(B12:G12;B2:G2). This formula calculates the sum of the products between the cells in line 12 (z) and line 2 (b_i);

– save your spreadsheet.

10.6.2. *Entering data*

To fill in our table, we will return to the example from Chapter 8 (section 8.3.2), which has three constraints and the following profit, z:

$$\begin{cases} 3x_1 + 1,5x_2 + 2x_3 \leq 6000 \\ 6x_1 + 4x_2 + 4x_3 \leq 10000 \\ x_1 + x_2 + x_3 \leq 3500 \end{cases}$$

$$z = 8x_1 + 3,5x_2 + 6x_3$$

Enter the data in your spreadsheet to obtain the table in Figure 10.38.

	A	B	C	D	E	F	G	H	I	J	K
1	Variables	x1	x2	x3	x4	x5	x6				
2	bi										
3											
4	Constraints										
5	n1	3	1,5	2				0	<=	6000	
6	n2	6	4	4				0	<=	10000	
7	n3	1	1	1				0	<=	3500	
8	n4							0	<=		
9	n5							0	<=		
10	n6							0	<=		
11											
12	z	8	3,5	6							
13											
14	Max z	0									
15											

Figure 10.38. *The data from the example, entered in our table*

NOTE.– The inequality signs (<=) in column I of the table are only there for information, they are not involved in the calculation.

10.6.3. *Implementing the solver*

Launch the solver, DATA tab, SOLVER tool in the ribbon. If it is not visible, consult Appendix 1 for the installation procedure.

Fill in the various solver fields:

– SET OBJECTIVE: $B$14;

– click on the option MAX (for this example, we are looking at maximization);

– BY CHANGING VARIABLE CELLS: $B$2:$G$2;

– SUBJECT TO THE CONSTRAINTS: $B$2:$G$2>=0; $H$10<=$I$10; $H$5<=$J$5; $H$6<=$J$6; $H$7<=$J$7; $H$8<=$J$8; $H$9<=$J$9;

NOTE.– In our example, all the constraints are not necessary but the solver parameters are set by default here so that it is able to resolve a problem comprising six inequalities.

– check the box: MAKE UNCONSTRAINED VARIABLES NON-NEGATIVE;

– SELECT A SOLVING METHOD: Simplex PL;

Figure 10.39. *The different parameters entered in the solver dialogue window*

– click the SOLVE button;

– after a moment, the results will be displayed in the table and the SOLVER RESULT window will be displayed.

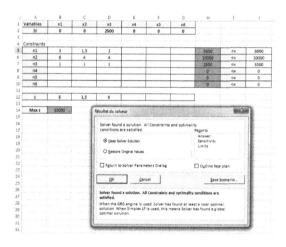

Figure 10.40. *All the calculated results and the SOLVER RESULT window*

The results found in Chapter 8 (section 8.3.7) are confirmed: $z = 15000$; $x1=0$; $x2=0$ and $x3 = 2500$.

Dashboards, Spreadsheets and Pivot Tables

11.1. The spreadsheet: a versatile tool

Since its arrival at the beginning of the 1970s, the spreadsheet has become a vital tool for business and moreover for any working environment. How many people use this tool to manage their bank accounts?

In this chapter, we will show you some of the possibilities associated with any spreadsheet worthy of the name, due to pivot tables, which provide excellent dashboards for aiding decision-making.

The examples that follow do not hinge on the spreadsheet's basic functions, but on advanced uses and on functions that are generally little known but which are very powerful in handling data and which can win precious time for decision-makers. We will, therefore, assume that the readers have sufficient experience to be able to create simple pivot tables.

In Chapter 10, we have already frequently used a (Microsoft Excel) spreadsheet, but here we are interested specifically in handling data from a database management system (DBMS) or a spreadsheet that itself constitutes a database for creating sophisticated dashboards or contingency tables.[1]

---

1 An idea introduced by the mathematician and statistician Karl Pearson in 1904, which consists of crossing two figures from a population and showing their combined result.

To carry out this work, we would use a 2013 version Microsoft Excel spreadsheet; however, the examples shown will be easily adaptable to other versions or other spreadsheets that include a form of pivot table such as, among others, "Data Pilot" in OpenOffice or DataPilot in LibreOffice.

11.2. The database: example

Before we begin our operations, we must create or import a database typical enough for our various operations to be coherent.

We suggest that you construct a base with several columns, grouped in a single table, created using a DBMS and then imported into Microsoft Excel or even entered directly into a spreadsheet.

The columns in our database will be:

– Date: date of sale, including day, month, year – format: dd/mm/yy;

– Designation: product label, "Product A" to "Product F" – format: text;

– Net UP: unit price before product tax – format: _($* #,##0.00_);

– Qty: quantity per item sold – format: whole number, 0;

– Vendor: vendor name – format: text;

– Agency: location of selling agency – format: text;

– Cat. client: client category, IND (Individual), PRO (Professional) – format: text;

– Client type: NW (New), EX (Existing) – format: text;

– Client residence zone: Ci (City), Di (District), C (County), S (State), OS (Outside state) – format: text;

– Net total: Quantity × Net unit price – format: _($* #,##0.00_).

In Figure 11.1, you can see a draft spreadsheet containing the database.

To achieve content with enough combinations to obtain interesting results, we strongly suggest you create a table containing at least 100 rows (tuples or records).

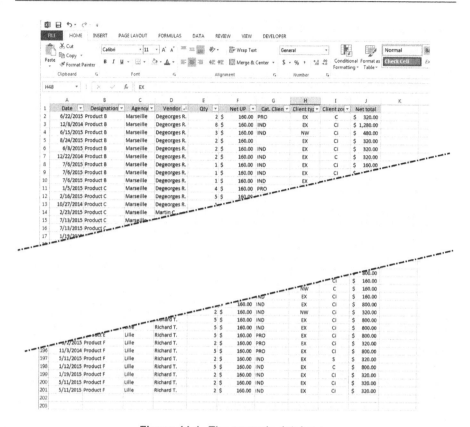

Figure 11.1. *The example database*

11.2.1. *Calculated field and formatting*

To begin, we would like to synthesize in a table, the net margin that each vendor has returned us for each product as a function of the client category. This margin is estimated as being 32% of the total net of a sale, whatever the product in question.

Moreover, it would be desirable for the purposes of interpretation, for us to be able to visualize the significance of the margin very quickly in graph form.

Figure 11.2 shows the result to be obtained.

M	N	O	P	Q	R	S	T
Client category	(All)						
Net margins	Designation						
VENDOR	Product A	Product B	Product C	Product D	Product E	Product F	Grand Total
Degeorges R.	$ -	$1,126.40	$ 512.00	$ 307.20	$ 307.20	$ 460.80	$ 2,713.60
Dupont V.	$ 204.80	$ 614.40	$ 460.80	$ 716.80	$ 256.00	$ 307.20	$ 2,560.00
Durant V.	$ 768.00	$ 204.80	$1,433.60	$ 51.20	$ 102.40	$ 512.00	$ 3,072.00
Fabre D.	$ 358.40	$ 256.00	$ -	$ 204.80	$1,075.20	$ 358.40	$ 2,252.80
Felin G.	$ 512.00	$ 460.80	$ -	$ 256.00	$ 768.00	$ -	$ 1,996.80
Marchand M.	$ 716.80	$ -	$ 204.80	$1,126.40	$ 409.60	$ 716.80	$ 3,174.40
Martin C.	$ 614.40	$1,024.00	$1,280.00	$1,075.20	$ 870.40	$ 460.80	$ 5,324.80
Pizarelli B.	$ 1,024.00	$ 102.40	$ 256.00	$1,433.60	$1,280.00	$ 716.80	$ 4,812.80
Richard T.	$ 153.60	$ 204.80	$ 409.60	$ 614.40	$ 512.00	$1,024.00	$ 2,918.40
Grand Total	$ 4,352.00	$3,993.60	$4,556.80	$5,785.60	$5,580.80	$4,556.80	$28,825.60

Figure 11.2. *The spreadsheet to be created. For a color version of the figure, see www.iste.co.uk/reveillac/logistics.zip*

We are going to create a simple table that we can then customize:

– create a pivot table with the row, "Vendor" and the column, "Designation";

– uncheck Autofit column widths on update in options;

– rename the headers to obtain a coherent table;

– resize the columns so that the product columns are identical in width;

– set all cells to currency format (Category: Accounting, Symbol: $ and Decimal places: 2).

	Designation						
VENDOR	Product A	Product B	Product C	Product D	Product E	Product F	Grand Total
Degeorges R.							
Dupont V.							
Durant V.							
Fabre D.							
Felin G.							
Marchand M.							
Martin C.							
Pizarelli B.							
Richard T.							
Grand Total							

Figure 11.3. *The initial table*

In order to display the margin total, we are going to have to insert a calculated field in our pivot table:

– select a cell in the pivot table, at the intersection of a seller and a product;

– place the cursor in the ANALYZE tab, and then choose the icon FIELDS, ITEMS AND SETS in the ribbon;

– select the option CALCULATED FIELD in order to open the dialogue box INSERT CALCULATED FIELD;

– in the NAME field, enter "Net margin";

– enter the FORMULA field, if there is a value 0, clear it and keep the = sign;

– in the list of fields, double-click on the field "Net total", it should appear beside the = sign;

– type"*0.32" (in order to calculate the 32%margin).

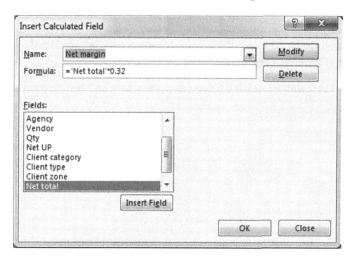

Figure 11.4. *Inserting the calculated field: Net total × 0.32*

– click on the ADD button, and then on OK;

– change the name "Net margin totals" (above left in the table) to "Net margins";

– your table should then fill with the new calculation.

Net margins	Designation ▼						
VENDOR ▼	Product A	Product B	Product C	Product D	Product E	Product F	Grand Total
Degeorges R.	$ -	$1,126.40	$ 512.00	$ 307.20	$ 307.20	$ 460.80	$ 2,713.60
Dupont V.	$ 204.80	$ 614.40	$ 460.80	$ 716.80	$ 256.00	$ 307.20	$ 2,560.00
Durant V.	$ 768.00	$ 204.80	$1,433.60	$ 51.20	$ 102.40	$ 512.00	$ 3,072.00
Fabre D.	$ 358.40	$ 256.00	$ -	$ 204.80	$1,075.20	$ 358.40	$ 2,252.80
Felin G.	$ 512.00	$ 460.80	$ -	$ 256.00	$ 768.00	$ -	$ 1,996.80
Marchand M.	$ 716.80	$ -	$ 204.80	$1,126.40	$ 409.60	$ 716.80	$ 3,174.40
Martin C.	$ 614.40	$1,024.00	$1,280.00	$1,075.20	$ 870.40	$ 460.80	$ 5,324.80
Pizarelli B.	$ 1,024.00	$ 102.40	$ 256.00	$1,433.60	$1,280.00	$ 716.80	$ 4,812.80
Richard T.	$ 153.60	$ 204.80	$ 409.60	$ 614.40	$ 512.00	$1,024.00	$ 2,918.40
Grand Total	$ 4,352.00	$3,993.60	$4,556.80	$5,785.60	$5,580.80	$4,556.80	$28,825.60

Figure 11.5. *The pivot table with calculation margins*

It only remains for us to include a histogram at the bottom, showing the value of each of the margins and that of their totals:

– select the margins contained in the column for "Product A";

– go to the HOME tab;

– choose the icon CONDITIONAL FORMATTING and then the option DATA BARS;

– select a font and a color. The column will be replaced by bars of proportional length to the values;

– repeat the operation for all the other products and for "Grand total";

– go to the HOME tab;

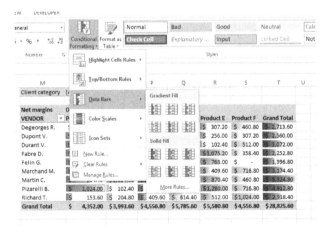

Figure 11.6. *The conditional formatting with its options. For a color version of the figure, see www.iste.co.uk/reveillac/logistics.zip*

– choose the icon FORMAT AS TABLE;

– select a format.

NOTE.– During the operations within the pivot tables, the column width adapts automatically to the contents by default, which can be distracting. To disable this option, you should simply select the ANALYZE tab, click on the OPTIONS icon in the PIVOTTABLE OPTIONS area in the ribbon, and then in the dialogue box that opens, in the tab LAYOUT & FORMAT, uncheck AutoFit....

Figure 11.7. *Setting the width parameters in pivot table options*

11.2.2. *Calendar sections, rankings and averages*

In this second example, we would like, with the help of a table bringing together agencies and products, to show the net averages of the quarterly sales of each establishment as a function of the client category and area of residence. A graph should highlight the best results; the agencies will be arranged in alphabetical order and the quarters by year, in increasing order.

Figure 11.8 shows the expected results.

| Cat. Client | (All) | | | | | | |
| Client zone | (All) | | | | | | |

| AVERAGE NET SALES | Designation | | | | | | |
Agency	Product A	Product B	Product C	Product D	Product E	Product F	Grand Total
⊟Bordeaux	$ 600.00		$ 85.00	$ 575.00	$ 2,480.00	$ 320.00	$ 601.05
⊟2014							
Qtr4	$ 720.00			$ 562.50			
⊟2015							
Qtr1	$ 120.00		$ 85.00	$ 450.00	$ 2,4...		
Qtr2	$ 840.00			$ 600.00			
Qtr3							$ 823.33
⊟Caen	$ 210.00	$ 241.67					
⊟2014						$ 480.00	$ 325.00
Qtr4					$ 1,550.00	$ 320.00	$ 661.88
⊟2015							$ 290.00
Qtr1				$ 562.50	$ 1,240.00	$ 480.00	$ 451.07
Qtr2							
Qtr3		...3.75	$ 255.00	$ 225.00	$ 620.00		$ 296.25
	$ 240.00	$ 193.33	$ 170.00	$ 1,350.00	$ 1,550.00	$ 160.00	$ 506.25
	$ 240.00	$ 217.50	$ 425.00	$ 450.00	$ 1,550.00	$ 480.00	$ 491.82
						$ 800.00	$ 800.00
...seille		$ 354.44	$ 311.67	$ 675.00	$ 852.50	$ 480.00	$ 469.17
⊞Nancy	$ 450.00	$ 290.00	$ 264.44	$ 375.00	$ 930.00	$ 533.33	$ 448.20
⊞Nice	$ 240.00	$ 326.25		$ 375.00	$ 1,162.50		$ 517.50
⊟Paris	$ 205.71	$ 362.50	$ 552.50	$ 787.50	$ 744.00	$ 480.00	$ 494.52
⊟2014							
Qtr4	$ 240.00	$ 435.00			$ 465.00	$ 320.00	$ 355.00
⊟2015							
Qtr1	$ 200.00	$ 290.00		$ 600.00		$ 320.00	$ 359.00
Qtr2	$ 120.00	$ 386.67	$ 680.00	$ 975.00	$ 930.00		$ 697.73
Qtr3			$ 425.00			$ 800.00	$ 612.50
Grand Total	$ 329.03	$ 305.68	$ 260.86	$ 605.36	$ 1,090.00	$ 474.67	$ 612.65

Figure 11.8. *The pivot table to be obtained*

As in the previous example, we will begin by creating the base table containing the following elements:

– FILTERS: "Client category" and "Client zone";

– COLUMNS: "Designation";

– ROWS: "Agency" and "Date";

– Σ VALUES: "Net total amount".

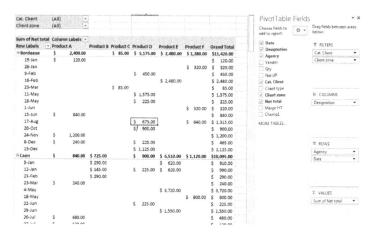

Figure 11.9. *The initial table for the second example*

Now, we are going to bring about the changes necessary to respond to demand:

– place the cursor over one of the sums, and then by right-clicking, select VALUE FIELD SETTINGS…;

– in the CUSTOM NAME field in the dialogue box that opens, enter "Average net sales";

– in the drop-down list, select AVERAGE;

– click on the NUMBER FORMAT button, and then in the dialogue box that opens, choose ACCOUNTING in the CATEGORY list, and then click on the OK button;

– click again on the OK button to confirm your choices and close the window. All the values in the table should be recalculated;

– in the table (above left), replace "Average of Net total" by "AVERAGE NET SALES";

– select one of the "day" dates, right-click and choose GROUP. In the drop-down list in the window that opens, choose QUARTERS and YEARS, and then click on the OK button;

– select all the cells under "Grand total", except the one on the last line;

– go to the HOME tab, click on the CONDITIONAL FORMATTING icon, select DATA BARS and choose a model;

– for greater clarity, you can also choose a format by clicking on the icon FORMAT AS TABLE in the STYLE group in the ribbon and choose a model there as well.

11.2.3. *Conditional calculated fields*

We wish to calculate the total premium of each of our vendors as a function of the quarter and month for each year chosen. The total premium is a percentage of the total monthly amount of net sales made in the course of the month. It follows the following rule:

– net total greater than or equal to $ 1,600.00: premium = 1.6%;

– net total greater than or equal to $ 2,500.00: premium = 2.2%.

The final table should resemble that in Figure 11.10.

Vendor	Total net sales		Net premium	
⊟ Degeorges R.				
⊟ 2014				
Oct	$	480.00	$	-
Dec	$	1,600.00	$	25.60
⊟ 2015				
Jan	$	1,440.00	$	-
Feb	$	1,120.00	$	-
May	$	960.00	$	-
Jun	$	1,120.00	$	
Jul	$	1,440.00		
Aug	$			
⊟ Dupont				
				25.60
			$	-
		760.00	$	28.16
	$	1,600.00	$	25.60
Aug	$	1,920.00	$	30.72
⊟ Richard T.				
⊟ 2014				
Nov	$	1,600.00	$	25.60
Dec	$	160.00	$	-
⊟ 2015				
Jan	$	2,240.00	$	35.84
Mar	$	640.00	$	-
May	$	1,120.00	$	-
Jun	$	2,720.00	$	59.84
Aug	$	640.00	$	-

Figure 11.10. *The pivot table with the total monthly premiums displayed*

To begin, we are going to create a table containing all the necessary items available:

– create a pivot table with "Vendor" and "Date" in the ROWS area;

– insert "Net total" in the Σ VALUES area;

– select GROUP by right-clicking on one of the dates in the table;

– choose "Months" and "Years" in the drop-down list and click on OK. The column should contain a list of months for each year, for each vendor;

– rename the column "Total net amount" as "Total net sales", and then change it to currency format;

Vendor	Total net sales
⊟Degeorges R.	$ 8,480.00
⊟2014	
Oct	$ 480.00
Dec	$ 1,600.00
⊟2015	
Jan	$ 1,440.00
Feb	$ 1,120.00
May	$ 960.00
Jun	$ 1,120.00
Jul	$ 1,440.00
Aug	$ 320.00
⊟Dupont V.	$ 8,000.00
⊟2014	
Oct	$ 960.00
Nov	$ 320.00
Dec	$ 160.00
⊟2015	
Jan	$ 1,440.00
Mar	$ 640.00
May	$ 2,880.00

Figure 11.11. *The pivot table that will serve as a base*

– create a calculated field (from a pivot table cell, ANALYZE tab, FIELDS, ITEMS AND SETS) named "Premium" and allocate it the following conditional formula:

```
=IF('Net Total'>=2500,2.2%*'Net
Total',IF('Net'>=1600,1.6%*'Net Total',0))
```

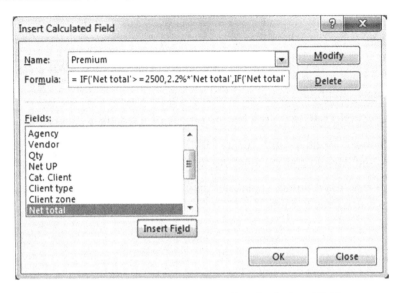

Figure 11.12. *Creating the conditional calculated field*

– drag the "Premium" field into the Σ VALUES area;

– rename the column "Net premium";

– you will be able to see that the subtotals per vendor are incorrect, as the pivot table tool is not capable of calculating them, we will delete them;

– place the cursor in one of the pivot table cells, go to the DESIGN tab, click on the icon SUBTOTALS shown in the ribbon and check DONOT SHOW SUBTOTALS;

– the same goes for the grand totals situated at the bottom of the table. Go to the ANALYZE tab, click on the OPTIONS icon in the PIVOT TABLE OPTIONS area in the ribbon, and then, in the dialogue box that opens, under the tab TOTALS & FILTERS, uncheck SHOW GRAND TOTALS FOR COLUMNS;

– click on the OK button to confirm.

11.2.4. *Slicers, filtering and calculated fields*

We have just seen several examples using a calculated field, now we are going to link the filter, calculated field and slicer.

In the database that we are using, in 2015 each of the vendors created a turnover, for products A to F, varying between $ 4,725.00 and $ 11,520.00.

We would like our pivot table to show us the target figure that the six least successful vendors should reach in 2016, knowing that this should be 12% higher than in 2015.

To begin, we will start with the very simple pivot table in Figure 11.13.

| Net total | Years | | |
Vendor ▾	2014	2015	Grand Total
Degeorges R.	$ 1,855.00	$ 6,835.00	$ 8,690.00
Dupont V.	$ 1,065.00	$ 7,580.00	$ 8,645.00
Durant V.	$ 1,905.00	$ 5,300.00	$ 7,205.00
Fabre D.	$ 770.00	$ 9,325.00	$ 10,095.00
Felin G.	$ 3,555.00	$ 4,725.00	$ 8,280.00
Marchand M.	$ 6,140.00	$ 11,520.00	$ 17,660.00
Martin C.	$ 2,840.00	$ 9,110.00	$ 11,950.00
Pernot C.		$ 8,095.00	$ 8,095.00
Picard D.	$ 2,205.00	$ 5,780.00	$ 7,985.00
Pizarelli B.	$ 3,690.00	$ 7,730.00	$ 11,420.00
Richard T.	$ 2,470.00	$ 8,150.00	$ 10,620.00
Grand Total	$ 26,495.00	$ 84,150.00	$ 110,645.00

Figure 11.13. *The initial pivot table, which will be modified*

The procedures below will delete the column for 2014, display the six least successful vendors of 2015, and then add a column calculating their targets for 2016:

– select one of the cells in the pivot table;

– go to the ANALYZE tab and in the ribbon, click the icon INSERT SLICER;

– in the dialogue box that opens, check "Date", and then click on the OK button;

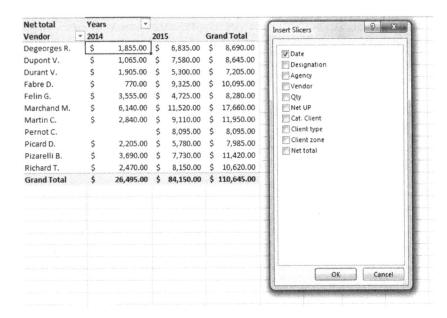

Figure 11.14. *Choosing a field on which to apply the slicer*

– in the filter interface that opens, double-click on 2015.

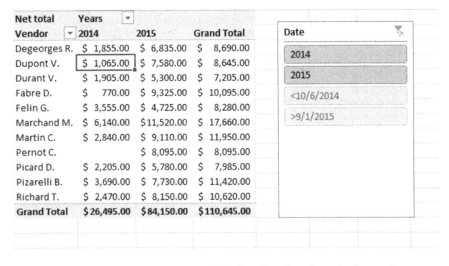

Figure 11.15. *The pivot table and the filter interface for selecting a slicer*

NOTE.– The filter interface for slicing remains available and you can change your options at any time by clicking on one of the relevant filter buttons.

If you do not wish to use the filter, which you ought to keep, it is possible to delete the year 2014 from the pivot table by using the slicer management option.

The step is, therefore, as follows (replacing the previous actions):

– select the cell 2014 in the pivot table;

– right click and choose FILTER, and then DATE FILTERS…;

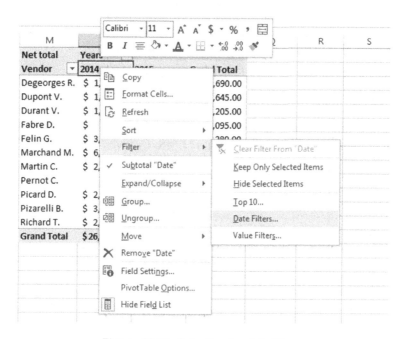

Figure 11.16. *Selecting the date filters*

– in the dialogue box that opens, choose IS AFTER in the drop-down list, and then enter 01/01/2015 in the field;

– click on the OK button. Your pivot table contains only the year 2015.

Net total	Years	$\blacktriangledown$
Vendor $\blacktriangledown$	2015	Grand Total
Degeorges R.	$ 6,835.00	$ 6,835.00
Dupont V.	$ 7,580.00	$ 7,580.00
Durant V.	$ 5,300.00	$ 5,300.00
Fabre D.	$ 9,325.00	$ 9,325.00
Felin G.	$ 4,725.00	$ 4,725.00
Marchand M.	$11,520.00	$11,520.00
Martin C.	$ 9,110.00	$ 9,110.00
Pernot C.	$ 8,095.00	$ 8,095.00
Picard D.	$ 5,780.00	$ 5,780.00
Pizarelli B.	$ 7,730.00	$ 7,730.00
Richard T.	$ 8,150.00	$ 8,150.00
Grand Total	$84,150.00	$84,150.00

Figure 11.17. *The pivot table contains only the year 2015*

We will now retain only the six least successful vendors and create our calculated item, the target figure to be reached in 2016:

– select one of the vendors;

– right click and choose FILTER, and then TOP 10;

– in the dialogue box that opens, with the help of the various drop-down lists, set the parameters: BOTTOM; 6; ITEMS; by "Net total";

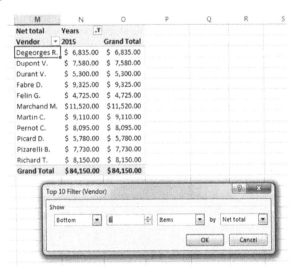

Figure 11.18. *The filter for selecting the six least successful vendors*

– your pivot table should only contain the six least successful vendors;

Net total	Years	
Vendor	2015	Grand Total
Degeorges R.	$ 6,835.00	$ 6,835.00
Dupont V.	$ 7,580.00	$ 7,580.00
Durant V.	$ 5,300.00	$ 5,300.00
Felin G.	$ 4,725.00	$ 4,725.00
Picard D.	$ 5,780.00	$ 5,780.00
Pizarelli B.	$ 7,730.00	$ 7,730.00
Grand Total	$37,950.00	$37,950.00

Figure 11.19. *The pivot table with the six least successful vendors*

– go to the ANALYZE tab, click on the icon FIELDS, ITEMS AND SETS in the ribbon and choose CALCULATED FIELD…;

– in the NAME field, enter "2016 target" ;

– in the FORMULA field, clear the value 0, then double-click on "Net total" in the drop-down list and then add "*1.12" to calculate the 12% increase required by the target;

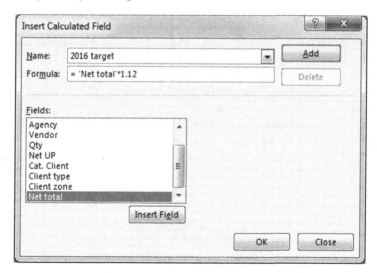

Figure 11.20. *Setting the parameters for the calculated field for the 2016 target*

– two new columns will appear in your pivot table, "2016 target amount" and "Total 2016 target amount";

Vendor	2015 Net total	Sum of 2016 target	Total Net total	Total Sum of 2016 target
Degeorges R.	$ 6,835.00	$ 7,655.20	$ 6,835.00	$ 7,655.20
Dupont V.	$ 7,580.00	$ 8,489.60	$ 7,580.00	$ 8,489.60
Durant V.	$ 5,300.00	$ 5,936.00	$ 5,300.00	$ 5,936.00
Felin G.	$ 4,725.00	$ 5,292.00	$ 4,725.00	$ 5,292.00
Picard D.	$ 5,780.00	$ 6,473.60	$ 5,780.00	$ 6,473.60
Pizarelli B.	$ 7,730.00	$ 8,657.60	$ 7,730.00	$ 8,657.60
Grand Total	$ 37,950.00	$ 42,504.00	$ 37,950.00	$ 42,504.00

Figure 11.21. *The two columns created by adding the field"2016target"*

– rename the first as "Target for 2016";

– select the ANALYZE tab, click on the OPTIONS icon in the area PIVOTABLE OPTIONS in the ribbon, and then in the dialogue box that opens, under the tab TOTALS & FILTERS, uncheck SHOW GRAND TOTALS FOR ROWS;

– click on the OK button to confirm.

The table that you should have obtained should be similar to that in Figure 11.22.

Vendor	2015 Net total	Target for 2016
Degeorges R.	$ 6,835.00	$ 7,655.20
Dupont V.	$ 7,580.00	$ 8,489.60
Durant V.	$ 5,300.00	$ 5,936.00
Felin G.	$ 4,725.00	$ 5,292.00
Picard D.	$ 5,780.00	$ 6,473.60
Pizarelli B.	$ 7,730.00	$ 8,657.60
Grand Total	$ 37,950.00	$ 42,504.00

Figure 11.22. *The finished pivot table with its column "Net total" for 2015 and the targets to be reached by the vendors in 2016*

11.2.5. *Calculated items*

Earlier in this work, we often used calculated fields, however, there is another option that limits the extent of the target – the calculated items. In this case, the calculation does not apply to all the items available in the database columns, but only to some.

Looking at a pivot table already made, which recapitulates on the year 2015 and crosses products with agencies, we can see that "Nice", "Bordeaux" and "Caen", all products combined, have made fewer sales than the other agencies.

Qty 2015 Designation ▾	Agency ▾ Bordeaux	Caen	Dijon	Lille	Lyon	Marseille	Nancy	Nice	Paris	Grand Total
Product A	8	7	14	2	4		15	2	6	58
Product B		5		4	7	12	4	8	14	54
Product C	1		2	8	7	21	15		13	67
Product D	13	2	22	12	14	6	1	1	21	92
Product E	8	21	8	5	10	11	12	10	9	94
Product F	8	5	14	15	12	7	5		7	73
Grand Total	38	40	60	46	54	57	52	21	70	438

Figure 11.23. *Quantities sold by each agency in 2015*

In order to balance this situation, we prognosticate an increase in the number of sales:

– 50% more for the agency in Nice;

– 40% more for the agencies in Bordeaux and Caen.

So that our vendors in these agencies have a better picture of what they should sell in 2016, we are going to create the pivot table from Figure 11.24, starting from the recapitulative pivot table above (Figure 11.23):

Qty 2015 Designation ▾	Agency ▾T Bordeaux	Bordeaux 2016	Caen	Caen 2016	Nice	Nice 2016
Product A	8	12	7	10	2	3
Product B		0	5	7	8	12
Product C	1	2		0		0
Product D	13	19	2	3	1	2
Product E	8	12	21	30	10	15
Product F	8	12	5	7		0
Grand Total	38	57	40	57	21	32

Figure 11.24. *The pivot table you should obtain with the columns Bordeaux, Caen and Nice 2016*

– in the pivot table, open the pop-up menu in the cell containing "Agency" by clicking on the filter icon and ensure that only the agencies for Nice, Caen and Bordeaux are checked;

– click on one of the agency names to select it;

– go to the ANALYZE tab, click on the icon FIELDS, ITEMS AND SETS in the ribbon and choose CALCULATED ITEM...;

– in the NAME field, enter "Nice 2016";

– place the cursor in the FORMULA field, clear the value 0, enter "CEILING" (behind the equal sign, then click on "Agency" in the drop-down list of fields, then double-click on "Nice" in the drop-down list of items and finally add "*1.5,1)" to calculate the 50% increases;

NOTE.– CEILING exists to round the product up to the nearest unit. The number 1 placed behind decimal point gives the rounded value.

Figure 11.25. *Creating the column that will contain the calculated item "Nice 2016". We can see the calculation formula for a 50% increase, with the function "CEILING"*

– click on the OK button to confirm creation of your new column;

– repeat the previous two operations with the agencies from Bordeaux and Caen, while entering a coefficient of 1.4 corresponding to 40%;

– select the ANALYZE tab, click on the OPTIONS icon in the PIVOT TABLE OPTIONS area in the ribbon and then in the dialogue box that opens, under the TOTALS AND FILTERS tab, uncheck SHOW GRAND TOTALS FOR ROWS;

– click on the OK button to confirm;

– open the pop-up menu from the cell containing "Agency" by clicking on the filter icon, and then click on the option SORT A TOZ. This tool is designed to place the columns for each city in order.

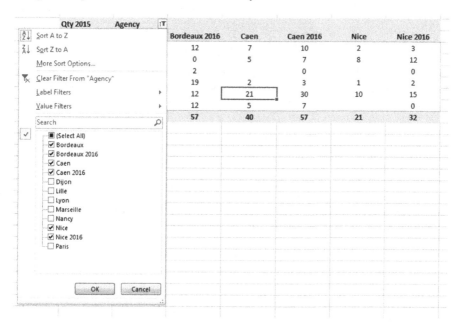

Figure 11.26. *The pop-up menu for the agencies with, among others, its sort options*

11.3. Multiple databases

One question that often arises when using pivot tables is the possibility of extracting data from several tables at once.

By definition, in a Microsoft Excel spreadsheet, it is only possible to have a single database defined on a range of cells containing a heading that describes each of the fields (columns) in the database.

This was the type of database that we used in the previous examples in this chapter.

There is, however, a very useful solution, which is little known even though a very powerful version has been developed in Microsoft Excel 2013. This is the table manager (for tables or data tables).

A range of cells can be transformed into a data table. We will not detail all the possibilities that this method of working presents here. If you wish to know more, consult the bibliography and internet links at the end of this work.

NOTE.– Some of the operations that follow are only possible with the 2013 version of Microsoft Excel. To use them in the 2010 version, you will have to download an extension module (addin) called "PowerPivot", downloadable on the Microsoft Website.

If you are using other versions of Excel or other spreadsheets, you should be aware that the possibilities presented will not necessarily be available.

To demonstrate to you how to use pivot tables linked to tables via one example, we will have to modify our initial database (see section 11.2).

We will confine ourselves to explaining that the table manager is in fact an extrapolation of the technique used in **relational database management systems (RDBMS)** in which tables[2] are linked to one another by **relationships** (links) with specific characteristics (**cardinalities**), as found in Microsoft Access, Oracle, IBM DB2, Sybase, etc. (see the bibliography at the end of this work).

11.3.1. *New tables for the database*

Our database will have two tables, "Sales" and "Products". The first is a modified version of the one that we have been using until now.

---

2 The vocabulary, "tables", "relationships", "requests" and "cardinalities", is used to exploit and manipulate databases that use what is commonly known as relational algebra. In the bibliography, you will be able to find works and Internet links that focus on these themes.

Table "Sales":

– Date: Date of sale with day, month, year – format: dd/mm/yy;

– Reference: FF NNN (family, n°) – format: whole number, 00 000;

– Agency: Location of selling agency – format: text;

– Vendor: Name of vendor – format: text;

– Qty: Quantity as a number of items – format: whole number, 0;

– Client category: IND (Individual), PRO (Professional) – format: text;

– Client type: NW (New), EX (Existing) – format: text;

– Client residence zone: Ci (City), Di (District), C (County), S (State), OS (Outside state) – format: text.

	A	B	C	D	E	F	G	H
1	Date	Reference	Agency	Vendor	Qty	Client cat.	Client type	Client zone
2	06/22/15	01 001	Marseille	Degeorges R.	2	PRO	EX	C
3	12/08/14	01 002	Marseille	Degeorges R.	8	IND	EX	Ci
4	06/15/15	02 010	Marseille	Degeorges R.	3	IND	NW	Ci
5	08/24/15	02 011	Marseille	Degeorges R.	2		EX	Ci
6	06/08/15	02 015	Marseille	Degeorges R.	2	IND	EX	Ci
7	12/22/14	03 020	Marseille	Degeorges R.	2	IND	EX	C
8	07/06/15	01 001	Marseille	Degeorges R.	1	IND	EX	
9	07/06/15	01 002	Marseille	Degeorges R.	1	IND		
10	07/06/15	02 010	Marseille	D...				
11	01/05/15							
							EX	Ci
					5	IND	EX	Ci
				Richard T.	1	IND	EX	Ci
	...12/15	01 001	Lille	Richard T.	1	IND	NW	C
189	05/18/15	01 001	Lille	Richard T.	1	IND	EX	Ci
190	06/29/15	01 002	Lille	Richard T.	5	IND	EX	Ci
191	06/08/15	02 010	Lille	Richard T.	2	IND	NW	Ci
192	06/08/15	02 011	Lille	Richard T.	5	IND	EX	Ci
193	06/29/15	02 015	Lille	Richard T.	5	IND	EX	Ci
194	11/10/14	01 001	Lille	Richard T.	5	PRO	EX	Ci
195	03/02/15	01 002	Lille	Richard T.	2	PRO	EX	Ci
196	11/03/14	02 010	Lille	Richard T.	5	PRO	EX	Ci
197	05/11/15	02 011	Lille	Richard T.	2	IND	EX	S
198	01/12/15	02 015	Lille	Richard T.	5	IND	EX	C
199	01/19/15	01 001	Lille	Richard T.	2	IND	EX	Ci
200	05/11/15	01 002	Lille	Richard T.	2	IND	EX	Ci
201	05/11/15	02 010	Lille	Richard T.	2	IND	EX	Ci
202								
203								
204								

Figure 11.27. *An example of the cell range for the table "Sales"*

Table "Products":

– Reference: FF NNN (family, n°) – format: whole number, 00 000;

– Designation: Product label, "Product A" to "Product F"– format: text;

– Net UP: Unit price before tax – format: _($* #,##0.00_);

– Packaging: number of products per unit of sale – format: whole number, 0;

– Availability: number of days before sale (IM: immediate, nD: x days, nW: x weeks) – format: text.

J	K	L	M	N
Reference	Designation	Net UP	Packaging	Availability
01 001	Product A	$ 120.00	1	7D
01 002	Product B	$ 145.00	1	IM
02 010	Product C	$ 85.00	4	10D
02 011	Product D	$ 225.00	2	IM
02 015	Product E	$ 310.00	4	3D
03 020	Product F	$ 160.00	1	2W

Figure 11.28. *An example of the cell range for the table "Products"*

11.3.2. *Creating data tables*

We will now transform our two cell ranges into data tables:

– place the cursor in one of the cells in the cell range "Sales";

– select the INSERT tab;

– in the TABLES area in the ribbon, click on the TABLE icon;

– a dialogue box CREATE TABLE will open. Check that the range indicated covers the whole cell range and click on the OK button;

Figure 11.29. *Creating the table*

– do the same for the second cell range ("Products") ;

– select the DESIGN tab;

– in the PROPERTIES group, enter the table name: "Products" in place of the name created by default;

– after having selected a cell from the "Sales" table, do the same – rename it.

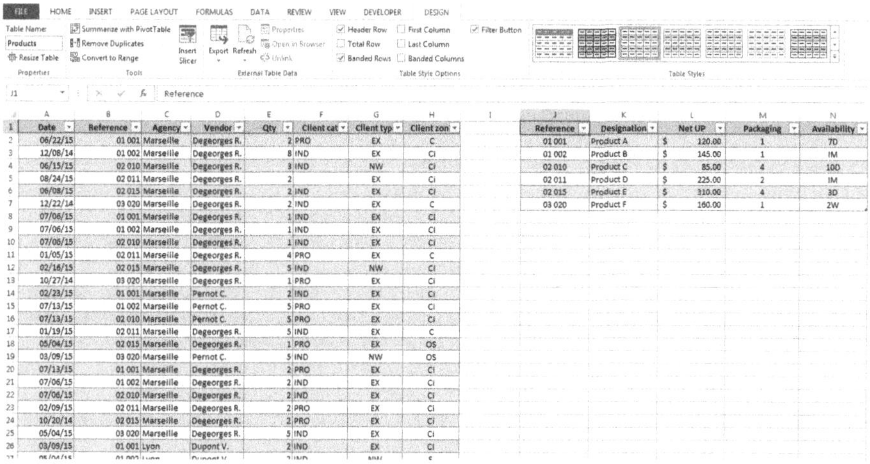

Figure 11.30. *The two tables, "Sales" (on the left) and "Products" (on the right)*

11.3.3. *Relationships between tables*

Now the tables have been created, we should link them (using a relationship or link) via a shared column, here "Reference".

Indeed, entering a product reference in the table "Sales" should enable us to know the information about it, whether its designation, net unit price, packaging or availability:

– place the cursor in one of the tables;

– select the DATA tab;

– in the DATA TOOLS group, click on the RELATIONSHIPS icon;

– a dialogue box opens, click on the NEW button;

– a new dialogue box CREATE RELATIONSHIP appears;

– in the drop-down list TABLE, choose "Sales";

– in the drop-down list RELATED TABLE, choose "Products";

– in the drop-down list COLUMN (FOREIGN), choose "Reference";

– in the drop-down list RELATED COLUMN (PRIMARY), choose "Reference";

NOTE.– The two tables can have columns (fields) with different names, they will still be linked on condition that they are of the same type. This is the same principal as for relational databases.

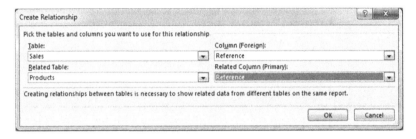

Figure 11.31. *The dialogue box for creating a relationship*

– click on the OK button;

– the dialogue box will close and your new relationship will appear in the MANAGE RELATIONSHIPS box;

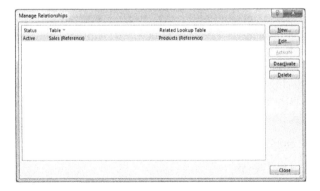

Figure 11.32. *The new relationship*

– click on the CLOSE button.

11.3.4. *Pivot tables using multiple tables*

Now that we have these two linked tables, we would like to display the pivot table from Figure 11.33.

Qty sold	Agency ▾									
	⊟Bordeaux	⊟Caen	⊟Dijon	⊟Lille	⊟Lyon		Picard D.	⊟Nice	⊟Paris	Grand Total
Designation ▾	Pizarelli B.	Fabre D.	Marchand M	Richard T.	Dupont V.			Felin G.	Martin C.	
Product A	16	10	20	12			11		2	25	111
Product B	5	4	13	10			3	1	11	19	93
Product C	6	6	11	9			6	3	5	8	74
Product D	23	10	7				10	10	6	13	112
Product E	7	7	9				19		13	13	110
Product F	3	7	2			5	11		2	9	63
Grand Total	60	44	62			17	60	14	39	87	563

Figure 11.33. *The pivot table to be created*

We can see that the quantities, vendors and agencies are derived from the "Sales" table, whereas the product designation comes from the "Products" table.

The relationship created previously establishes the link between the two tables with the help of the "Reference" column.

Here are all the operations to be carried out in order to create the dashboard summarizing the sales:

– place the cursor in the "Sales" table;

– select the INSERT tab;

– in the TABLES group, click on the icon PIVOT TABLE;

– in the dialogue box that opens, check that you have "Sales" in the TABLE/RANGE field;

– in LOCATION, enter the cell defining the location destination for your pivot table;

– check ADD THIS DATA TO THE DATA MODEL in order to take account of the two tables and their relationship;

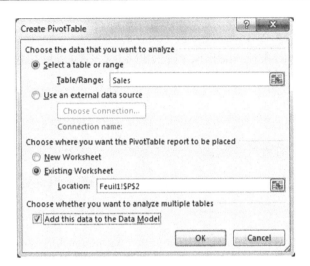

Figure 11.34. *Creating the pivot table using multiple tables*

– click on the OK button to confirm creation;

The pivot table fields area should resemble that in Figure 11.35. In the ACTIVE tab especially, you should see both your tables.

Figure 11.35. *The pivot table fields area with both active tables: "Products" and "Sales"*

– place the "Agency" and "Vendor" fields in the COLUMNS area;

– place the "Designation" field in the ROWS area;

– place the "Qty" field in the Σ VALUES area;

– rename the different items: table, row, header column and finally format the pivot table;

Qty sold	Agency ▾		
	⊟Bordeaux	Bordeaux Total	⊟Ca
Designation ▾	Pizarelli B.		Fabr
Product A	16	16	
Product B	5	5	
Product C	6	6	
Product D	23	23	
Product E	7	7	
Product F	3	3	
Grand Total	60	60	

Figure 11.36. *The renamed headers: "Qty sold", "Designation" and "Agency"*

– place the cursor in the pivot table, select the DESIGN tab;

– go to the LAY OUT group, click on the SUBTOTALS icon and choose DO NOT SHOW SUB TOTALS in order to make the subtotals for each of the agencies disappear.

11.4. Limits to and constraints on calculated fields

Let us now return to our first database, and imagine that we wish to delete the net total column, assuming that it is unnecessary since we could replace it by a calculated field within the pivot table.

The Net total column is the result of a calculation, row-by-row, multiplying product Quantity × Net unit price.

The final table should resemble that in Figure 11.37.

Designation ▾	Total qty	Net unit price		Net total	
Product A	85	$	120.00	$	10,200.00
Product B	78	$	145.00	$	11,310.00
Product C	89	$	85.00	$	7,565.00
Product D	113	$	225.00	$	25,425.00
Product E	109	$	310.00	$	33,790.00
Product F	89	$	160.00	$	14,240.00
Grand Total	563	$	177.05	$	102,530.00

Figure 11.37. *The table to be obtained*

In the following section, we will show you how to solve this while avoiding the traps created by the pivot table calculator.

On the surface, the solution to the problem seems simple.

Let us apply the following step:

– place the product "Designation" in the ROWS area;

– place the "Quantity" and the "Net unit price" in the Σ VALUES area;

– create a calculated field named "Net total", corresponding to the product of the "Quantity" multiplied by the "Net unit price";

– place the latter in the Σ VALUES area;

– in the table, replace the column headers by "Designation", "Total qty", "Net unit price" and "Net total";

– select all the numeric values in the table, go to the HOME tab and then choose CURRENCY in the drop-down list from the NUMBER group in the ribbon.

Although the step that we have applied seems logical, the results displayed are incorrect.

Designation ▾	Total qty	Net unit price	Net total
Product A	85	$ 3,720.00	$ 316,200.00
Product B	78	$ 5,365.00	$ 418,470.00
Product C	89	$ 2,465.00	$ 219,385.00
Product D	113	$ 9,450.00	$ 1,067,850.00
Product E	109	$ 9,610.00	$ 1,047,490.00
Product F	89	$ 4,800.00	$ 427,200.00
Grand Total	563	$ 35,410.00	$ 19,935,830.00

Figure 11.38. *The table resulting from our operations. The values obtained are inconsistent*

In fact, the pivot table calculator has not made any errors.

In the column "Net unit price", it simply adds the "Net unit price" of the product as many times as there are rows containing it.

In the column "Net total", it multiplies the "Net unit price" by the quantity. For example, for the first row: *85 x 3720.00 $= 316 200.00 $.*

It follows that the column totals are also inconsistent.

To overcome this problem, it is not, therefore, necessary to keep a total in the ΣVALUES area for the "Net unit price" but to change it to an average. The average of a group of equal values (here the unit price of a product) is identical to one of those values.

As for the calculated field, it is necessary to change the formula so that it calculates using the average of the net unit price.

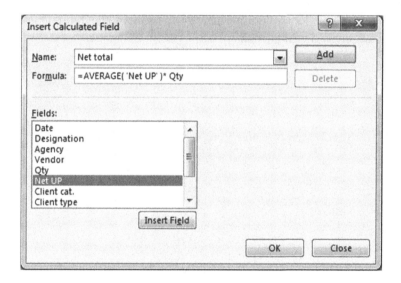

Figure 11.39. *The modified formula for calculating the "net total"*

This simple example shows the pivot table's limits in some scenarios. In principle, it is generally more useful, less risky and simpler to integrate a calculation by adding a column to the corresponding table in the database. The values contained in this column can then be manipulated very easily within a pivot table.

NOTE.– The solution that we put forward for this example will not function if the net unit price of a product varies over time, as the average would no longer be representative of the average over the whole period concerned.

11.5. Conclusion

In this chapter, we have shown some of the possibilities derived from this amazing tool the pivot table calculator within the Microsoft Excel spreadsheet.

We have omitted to mention many other possibilities and functions, whether linked to pivot tables or not, especially the functions of importing from databases, analyzing scenarios, validating and consolidating data, etc.

We invite you to explore further by consulting the bibliography and weblinks at the end of this book.

Scheduling and Planning with a Project Manager

12.1. Reminders and information

In this chapter, we are going to show you how a project is constructed using the manager Microsoft Project 2013. The objective is to introduce the possibilities presented by this type of software and its potential interaction with other tools.

Our intention is not to train you in all the functions that this manager offers, as a work such as this is not extensive enough.

The progress of this presentation will depend on the ideas seen in detail in Chapter 5. To these will be added resource implementation and management, awareness of the overallocation of resources and opportunities for reducing and leveling to best satisfy any constraints imposed, as well as monitoring the project.

All the work will be built on a concrete example taken from industry. Through this means, we hope to show you all the potential that the computing tools now offers in managing and monitoring one or more projects in a business.

Everything that will be explained below can easily be transcribed, partly or entirely, into project managers such as those mentioned in section 9.3.1.

12.2. An example: designing and making a machine-tool

12.2.1. *Scenario*

The MecaTools firm, which specializes in made-to-order machine tools, has to design a hydraulic press for stamping aluminum plate for one of its clients. To carry out this order, it draws up a list of specifications detailing the technical characteristics that the machine must possess.

The design, construction, tests and client approval will be carried out entirely in their workshops. To this end, the management has named a project leader who will be given the task of scheduling and planning the work to be carried out.

After having studied all the specifications required by the client, aided by a team of collaborators, the project leader has drawn up a precedence table that groups together all the main activities to be carried out.

Activity	Designation	Duration	Predecessor
A	Defining specifications	4 weeks	–
Design			
B	Basic design	3 weeks	A
C	Detailed design	4 weeks	B
D	Mechanical design	2 weeks	C
E	Hydraulic design	2 weeks	C
F	Automatic computer design	1 week	C
Constructing a prototype			
G	Mechanical construction	7 weeks	D
H	Assembling a hydraulic system	4 days	E, G
I	Assembling sensors and PLC	3 days	F, G
J	Wiring (electrical wiring, sensors, field network, etc.)	4 days	H, I
K	Developing and coding PLC application	4 weeks	F

L	Developing and coding computer application	5 weeks	F
M	Linking network to information systems	1 day	J, L
N	Integrating mechanics, hydraulics and automation	6 days	I, J, K
Tests and checks			
O	Network test	2 days	M, N
P	Testing and checking prototype	1 week	O, Q
Q	Technical documentation	2 weeks	C
Marketing			
R	Finalizing technical documentation	2 days	P, Q
S	Commercial documentation	1 week	P, Q
T	Presenting to client and approval	1 day	R, S

Table 12.1. *Tasks and predecessors*

The team of employees linked to this project is as follows. (The materials and raw materials used are not taken into account during planning, management of this is independently guaranteed by the MecaTools CAPM).

The hourly costs marked are the average hourly costs, inclusive of charges, for an employee or service.

Name	Function	Nbr	Hourly cost ($/h)	Timetable
Project manager	Leadership	1	$52.00	Timetable 1
Assistant mechanic for project	Production	1	$38.00	Timetable 1
Hydraulics assistant for project	Production	1	$38.00	Timetable 1
Computer automation assistant for project	Production	1	$41.00	Timetable 1
Design office (five people)	Production	1	$215.00	Timetable 2
Mechanical engineer	Production	1	$42.00	Timetable 3

Electrical engineer	Production	1	$44.00	Timetable 3
Hydraulic engineer	Production	1	$42.00	Timetable 3
Computer engineer	Production	1	$44.00	Timetable 3
Developer	IT	2	$35.50	Timetable 2
Fitter	Production	5	$18.00	Timetable 3
Technician	Production	4	$24.50	Timetable 3
Secretary-editor	Administration	1	$20.00	Timetable 4
Marketing (three people)	Administration	1	$160.00	Timetable 4
Commercial sector (three people)	Administration	1	$190.00	Timetable 4
Logistics (two people)	Administration	1	$158.00	Timetable 4
Quality control (three people)	Production	1	$144.00	Timetable 3
Team leader	Production	2	$30.00	Timetable 3

Table 12.2. *Resources*

Below, you will find the working hours for the business and personnel (they are different for different services):

– MecaTools factory hours: Monday to Saturday – 7:00 am to 8:00 pm (Annual closing: 07/25/2015 inclusive to 08/08/2015 inclusive);

– timetable 1: Monday to Friday – 8:00 am to 12:30 pm and 1:30 pm to 6:00 pm;

– timetable 2: Monday to Friday – 8:00 am to 12:00 pm and 1.30 pm to 5:30 pm;

– timetable 3: Monday to Friday– 8:00 am to 12:00 pm and 1:00 pm to 5:30 pm;

– timetable 4: Monday to Friday– 8:00 am to 12:00 pm and 2:00 pm to 5:30 pm.

The allocation of resources, overseen by the project leader, is given in Table 12.3.

Name	Allocation
Project manager	A, B, C, G, N, O,P, T
Assistant mechanic for project	A, B, D
Hydraulics assistant for project	A, B, E
Computer automation assistant for project	A, B, F
Design office (five people)	B, C, G
Mechanical engineer	B, C, D, G, N, P, T
Electrical engineer	B, C, F, I, J, K, N, P, T
Hydraulic engineer	B, C, E, H, N, P, T
Computer engineer	B, C, F, J, L, M, N, O, P, T
Developer 1	K, P
Developer 2	L, P
Fitter 1	H, M, N, P
Fitter 2	G, H, N
Fitter 3	N, M
Fitter 4	J, M
Fitter 5	I, J
Technician 1	G, J, N, O, P
Technician 2	G, J, N, O, Q
Technician 3	I, M, P
Technician 4	I, M, Q, R
Secretary-editor	A, B, C, Q, R
Marketing service (three people)	S
Commercial sector (three people)	A, S, T
Logistics (two people)	C, G, N
Quality control (three people)	N, P, Q, S, T
Team leader 1	H, N
Team leader 2	I, N

Table 12.3. *Allocation*

The start date for the project is fixed at 8:00 am on 06/01/2015.

12.2.2. *Designing and setting parameters for the project*

First, we will create the project calendar, i.e. the factory timetable, from Monday to Friday, 7:00 am to 8:00 pm.

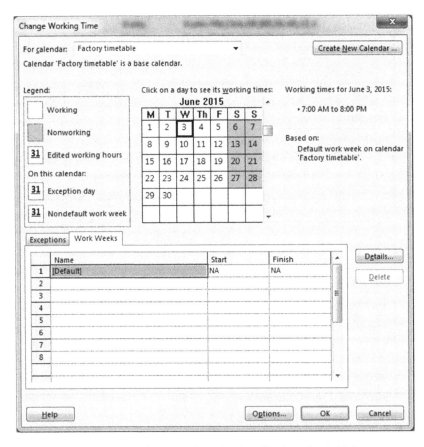

Figure 12.1. *Creating the calendar: "Factory timetable"*

In the calendar exceptions, we will place the annual holiday closing period, as well as public holidays (7/4/2015, 9/7/2015, 10/12/2015, 11/11/2015, 11/26/2015, 12/25/2015 and 01/01/16).

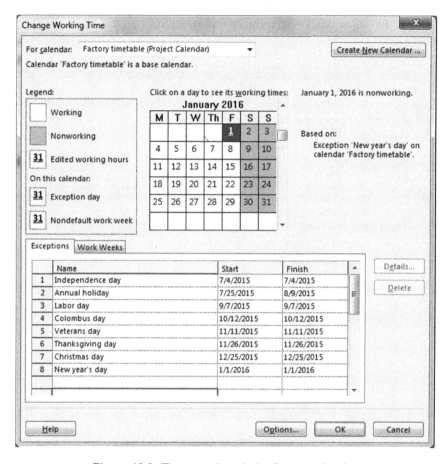

Figure 12.2. *The exceptions in the factory calendar*

Eventually, we will also change the options:

– the week commences on: Monday;

– default start time: 8:00 am;

– default finish time: 6:00 pm;

– hours per day: 7.5;

– hours per week: 37.5;

– days per month: 20.

Now, let us modify the project information, i.e. the start date and calendar.

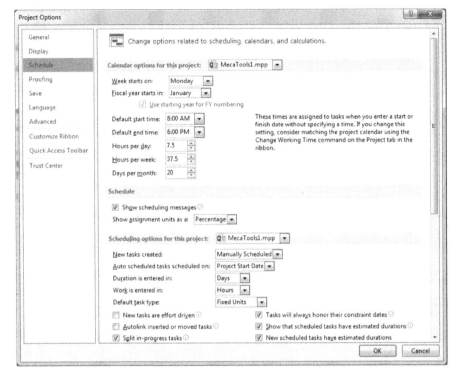

Figure 12.3. *Project calendar options*

Figure 12.4. *Project information*

Once these parameters have been entered, we will save the project under the name "MecaTools.mpp".

12.2.3. *Entering tasks and durations*

In addition to the tasks specified in the precedence table, we will add a task "Project launch", lasting 0 days, which will serve as an opening task as well as a task, "Finish project", also lasting 0 days, which will close it.

The four main types of summary task are defined as follows: "Design", "Prototype construction", "Tests and checks" and "Marketing" which each covers several classic tasks. To do this, we use the functions INDENT and OUTDENT TASKS, in the ribbon, under the TASK tab in the SCHEDULE group.

The planning mode for all the tasks is set at "automatic mode" (TASK tab, TASKS group, AUTO SCHEDULE icon in the ribbon).

The result obtained can be seen in Figure 12.5.

Figure 12.5. *All the main, summary and classic tasks entered in Microsoft Project*

It is striking that a task such as "design" has an automatically calculated duration of 20 days (a summary task) although it begins on 06/01/15 at

8:00 am and ends on 06/16/15 at 3:00 pm, which is slightly fewer than 16 calendar days.

The same goes for the other tasks, whose duration was entered by hand, for example the task "Assembling captors and PLC", lasting 3 days begins on 06/01/15 at 8 am and ends on 06/02/15 at 5:30 pm, which is a little under 2 calendar days.

In fact, there is no error, our project calendar is "Factory timetable" with a working day, see Figure 12.2, which begins at 7 am and ends at 8 pm, which is 13 hours long. If we return to our two tasks we can calculate, by looking at the default options in Figure 12.3, which shows 7.5 h per day:

– "Design": 20 days, so *(20 × 7.5)/13 ≈11.538d = 11d + 0.538d = 11d + 0.538 + 13 ≈11 d + 7 h* to which we must add 4 days off (2 Saturdays + 2 Sundays), so *11 d + 4 d + 7 h = 15 d + 7 h.* We can verify (with working days starting at 8:00 am): *06/01/15 8:00 am + 15 d 7 h = 06/16/15 3:00pm;*

– "Assembling sensors and PLC": 3 days, so (3 × 7.5)/13 ≈1.73 d = 1 w + 0.73 w = 1 w + 0.73 × 13 ≈1 w + 9.5 h. We can verify (with working days starting at 8:00 am): 06/01/15 8:00 am + 1 d 9.5 h = 06/02/15 5.30:00pm.

12.2.4. *Enter predecessors*

We can continue building our project by placing antecedents for each of our tasks.

We can add a Task column (Text1) to be able to note the relevant letters for the different tasks in order to facilitate entry of antecedents.

Finally, we obtain the table and Gantt chart from Figure 12.6.

12.2.5. *MPM network visualization*

The whole project can be visualized as an MPM network, using the NETWORK DIAGRAM in the ribbon, under the VIEW tab.

In this display mode, it is also possible to modify the network, especially by adding or deleting particular links.

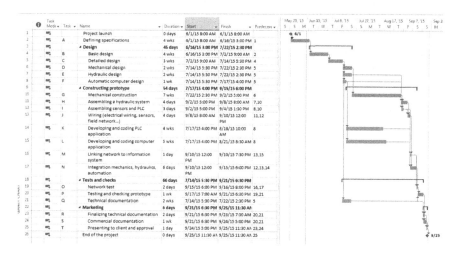

Figure 12.6. *The table with its completed columns "Task" and "Antecedents". For a color version of the figure, see www.iste.co.uk/reveillac/logistics.zip*

By using the zoom functions, we can see some parts of the network in detail. The data that are visible can vary; by default, we find the task name, its start date, end date, number, duration and resources, if they are allocated.

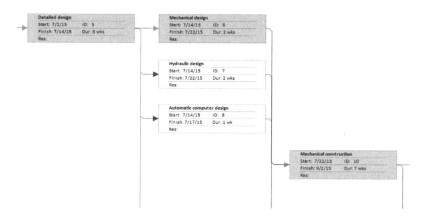

Figure 12.7. *Some "MecaTools" project tasks displayed in detail. For a color version of the figure, see www.iste.co.uk/reveillac/logistics.zip*

The dashed lines mark the edges of the pages (depending on the default printer settings) and thus indicate the page divisions that will occur during printing. In our example, the network will take up 15 A4 pages.

12.2.6. *Calculating slack*

With Microsoft Project, calculating slack is a mere formality, you simply need to complete the table to the left of the Gantt chart with the columns "Free slack" and "Total slack".

They are expressed by default in units similar to those used for duration.

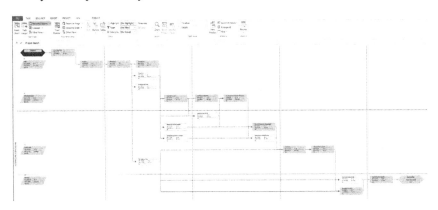

Figure 12.8. *The MPM network (or network of tasks) corresponding to the "MecaTools" project. For a color version of the figure, see www.iste.co.uk/reveillac/logistics.zip*

#	Task Mode	Task	Name	Duration	Start	Finish	Predecess	Free Slack	Total Slack
1			Project launch	0 days	6/1/15 8:00 AM	6/1/15 8:00 AM		0 days	0 days
2		A	Defining specifications	4 wks	6/1/15 8:00 AM	6/16/15 3:00 PM	1	0 wks	0 wks
3			▲ Design	45 days	6/16/15 3:00 PM	7/22/15 2:30 PM		0 days	0 days
4		B	Basic design	4 wks	6/16/15 3:00 PM	7/2/15 9:00 AM	2	0 wks	0 wks
5		C	Detailed design	3 wks	7/2/15 9:00 AM	7/14/15 5:30 PM	4	0 wks	0 wks
6		D	Mechanical design	2 wks	7/14/15 5:30 PM	7/22/15 2:30 PM	5	0 wks	0 wks
7		E	Hydraulic design	2 wks	7/14/15 5:30 PM	7/22/15 2:30 PM	5	7 wks	7 wks
8		F	Automatic computer design	1 wk	7/14/15 5:30 PM	7/17/15 4:00 PM	5	0 wks	0 wks
9			▲ Constructing prototype	54 days	7/17/15 4:00 PM	9/15/15 6:00 PM		0 days	0 days
10		G	Mechanical construction	7 wks	7/22/15 2:30 PM	9/2/15 5:00 PM	6	0 wks	0 wks
11		H	Assembling a hydraulic system	4 days	9/2/15 5:00 PM	9/8/15 8:00 AM	7,10	0 days	0 days
12		I	Assembling sensors and PLC	3 days	9/2/15 5:00 PM	9/4/15 1:30 PM	8,10	1 day	1 day
13		J	Wiring (electrical wiring, sensors, field network...)	4 days	9/8/15 8:00 AM	9/10/15 12:00 PM	11,12	0 days	0 days
14		K	Developing and coding PLC application	4 wks	7/17/15 4:00 PM	8/18/15 10:00 AM	8	5.6 wks	5.6 wks
15		L	Developing and coding computer application	5 wks	7/17/15 4:00 PM	8/21/15 8:30 AM	8	4.6 wks	5.6 wks
16		M	Linking network to information system	1 day	9/10/15 12:00 PM	9/10/15 7:30 PM	13,15	5 days	5 days
17		N	Integration mechanics, hydraulics, automation	6 days	9/10/15 12:00 PM	9/15/15 6:00 PM	12,13,14	0 days	0 days
18			▲ Tests and checks	66 days	7/14/15 5:30 PM	9/21/15 6:30 PM		0 days	0 days
19		O	Network test	2 days	9/15/15 6:00 PM	9/16/15 8:00 PM	16,17	0 days	0 days
20		P	Testing and checking prototype	1 wk	9/17/15 7:00 AM	9/22/15 6:30 PM	19,21	0 wks	0 wks
21		Q	Technical documentation	2 wks	7/14/15 5:30 PM	7/22/15 2:30 PM	5	10.2 wks	10.2 wks
22			▲ Marketing	6 days	9/21/15 6:30 PM	9/25/15 11:30 AM		0 days	0 days
23		R	Finalizing technical documentation	2 days	9/21/15 6:30 PM	9/23/15 7:30 AM	20,21	3 days	3 days
24		S	Commercial documentation	1 wk	9/21/15 6:30 PM	9/24/15 5:00 PM	20,21	0 wks	0 wks
25		T	Presenting to client and approval	1 day	9/24/15 5:00 PM	9/25/15 11:30 AM	23,24	0 days	0 days
26			End of the project	0 days	9/25/15 11:30 AM	9/25/15 11:30 AM	25	0 days	0 days

Figure 12.9. *The table, to the left of the Gantt chart, to which two columns have been added: "Free slack" and "Total slack". For a color version of the figure, see www.iste.co.uk/reveillac/logistics.zip*

It is also possible to display the slack on a Gantt chart by setting style bars.

Figure 12.10. *The Gantt chart with the free and total slack represented. Below, you can see the window for setting the style for the bars. For a color version of the figure, see www.iste.co.uk/reveillac/logistics.zip*

We will now progress to entering resources in the Microsoft Project resources table.

12.2.7. Entering resources

Before entering resources, we should create the calendars that we are going to need, "Timetable 1" to "Timetable 4" (see section 12.2.1).

To make it easier to create, we can make calendar copies from the "Factory timetable" calendar, in order to keep the holidays and annual break.

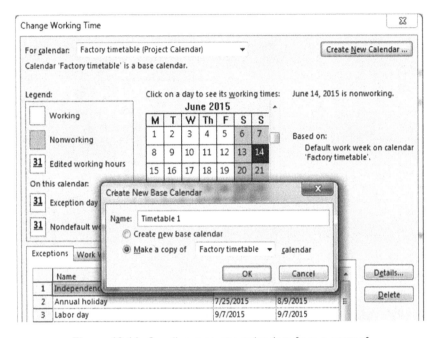

Figure 12.11. *Creating resource calendars from a copy of the "Factory timetable" calendar*

When several resources have the same function, we differentiate them and each of them is represented in a row.

By default, if no capacity is specified, each of the resources is considered as contributing 100% to this project.

The capacity value is often linked to the fact that a resource can be shared by several projects and can therefore have a capacity varying between 0 and 100% over the project in question.

No hourly rate for overtime has been entered since nothing has been specified in the list of charges.

As in most of the cases, when we look at the workforce, the allocations of cost are proportionate (to the time spent on the project).

In Figure 12.12, you will find the resource table.

	Resource name	Type	Material Label	Initials	Group	Capacit max.	Tx. standard	Tx. hrs. sup.	Coût/Uti	Accrue At	Base Calendar
1	Project manager	Work		PM		100%	$52.00/hr	$0.00/hr	$0.00	Prorated	Timetable 1
2	Assistant mechanic for project	Work		AM		100%	$38.00/hr	$0.00/hr	$0.00	Prorated	Timetable 1
3	Assistant hydraulic for project	Work		AH		100%	$38.00/hr	$0.00/hr	$0.00	Prorated	Timetable 1
4	Computer automation assistant for project	Work		CAA		100%	$41.00/hr	$0.00/hr	$0.00	Prorated	Timetable 1
5	Design office	Work		DO		100%	$215.00/hr	$0.00/hr	$0.00	Prorated	Timetable 2
6	Mechanical engineer	Work		ME		100%	$42.00/hr	$0.00/hr	$0.00	Prorated	Timetable 3
7	Electrical engineer	Work		EE		100%	$44.00/hr	$0.00/hr	$0.00	Prorated	Timetable 3
8	Hydraulic engineer	Work		HE		100%	$42.00/hr	$0.00/hr	$0.00	Prorated	Timetable 3
9	Computer engineer	Work		CE		100%	$44.00/hr	$0.00/hr	$0.00	Prorated	Timetable 3
10	Developer 1	Work		D1		100%	$35.50/hr	$0.00/hr	$0.00	Prorated	Timetable 2
11	Developer 2	Work		D2		100%	$35.50/hr	$0.00/hr	$0.00	Prorated	Timetable 2
12	Fitter 1	Work		F1		100%	$18.00/hr	$0.00/hr	$0.00	Prorated	Timetable 3
13	Fitter 2	Work		F2		100%	$18.00/hr	$0.00/hr	$0.00	Prorated	Timetable 3
14	Fitter 3	Work		F3		100%	$18.00/hr	$0.00/hr	$0.00	Prorated	Timetable 3
15	Fitter 4	Work		F4		100%	$18.00/hr	$0.00/hr	$0.00	Prorated	Timetable 3
16	Fitter 5	Work		F5		100%	$18.00/hr	$0.00/hr	$0.00	Prorated	Timetable 3
17	Technician 1	Work		T1		100%	$24.50/hr	$0.00/hr	$0.00	Prorated	Timetable 3
18	Technician 2	Work		T2		100%	$24.50/hr	$0.00/hr	$0.00	Prorated	Timetable 3
19	Technician 3	Work		T3		100%	$24.50/hr	$0.00/hr	$0.00	Prorated	Timetable 3
20	Technician 4	Work		T4		100%	$24.50/hr	$0.00/hr	$0.00	Prorated	Timetable 3
21	Secretary-editor	Work		SE		100%	$20.00/hr	$0.00/hr	$0.00	Prorated	Timetable 4
22	Marketing	Work		MK		100%	$160.00/hr	$0.00/hr	$0.00	Prorated	Timetable 4
23	Commercial sector	Work		CS		100%	$190.00/hr	$0.00/hr	$0.00	Prorated	Timetable 4
24	Logistics	Work		LO		100%	$158.00/hr	$0.00/hr	$0.00	Prorated	Timetable 4
25	Quality control	Work		QC		100%	$144.00/hr	$0.00/hr	$0.00	Prorated	Timetable 3
26	Team leader 1	Work		TL1		100%	$30.00/hr	$0.00/hr	$0.00	Prorated	Timetable 3
27	Team leader 2	Work		TL2		100%	$30.00/hr	$0.00/hr	$0.00	Prorated	Timetable 3

Figure 12.12. *The resource table*

Note that there are initials in this table which will be used a little later, when the resources are displayed on the Gantt chart.

12.2.8. *Allocation of resources*

We are not going to define which resources will be used in each of the tasks.

The software will automatically recalculate the data according to the constraints imposed (timetables, cost, slack, charges, potential overuse, etc.).

To carry out the allocations, we use the RESOURCE NAMES column in Gantt chart view.

The RESOURCE INITIALS column has been added and we ask it to be displayed behind each bar in the diagram.

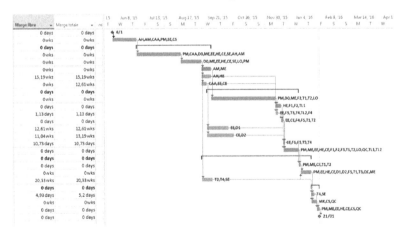

	ⓘ	Task Mode	Task	Name	Duration	Start	Finish	Predeces	Free Slack	Total Slack	Resource Names	Resource Initials
1				Project launch	0 days	6/1/15 8:00 AM	6/1/15 8:00 AM		0 days	0 days		
2			A	Defining specifications	5.33 wks	6/1/15 8:00 AM	7/2/15 10:30 AM	1	0 wks	0 wks	Assistant hydraulic for p	AH,AM,CAA,PM,SE,
3				◢ Design	99.73 days	7/2/15 10:30 AM	10/6/15 5:30 PM		0 days	0 days		
4			B	Basic design	7.79 wks	7/2/15 10:30 AM	8/28/15 11:00 AM	2	0 wks	0 wks	Project manager,Compu	PM,CAA,DO,ME,EE,I
5			C	Detailed design	4.41 wks	8/28/15 11:00 AM	9/24/15 9:30 AM	4	0 wks	0 wks	Design office,Mechanica	DO,ME,EE,HE,CE,SE,
6			D	Mechanical design	2.23 wks	9/24/15 9:30 AM	10/6/15 5:30 PM	5	0 wks	0 wks	Assistant mechanic for p	AM,ME
7			E	Hydraulic design	2.23 wks	9/24/15 9:30 AM	10/6/15 5:30 PM	5	18.89 wks	18.89 wks	Assistant hydraulic for p	AH,HE
8			F	Automatic computer design	1.21 wks	9/24/15 9:30 AM	10/1/15 8:30 AM	5	0 wks	16.17 wks	Computer automation a	CAA,EE,CE
9				◢ Constructing prototype	137.13 day	10/1/15 8:30 AM	1/27/16 10:00 AM		0 days	0 days		
10	◆		G	Mechanical construction	12.91 wks	10/6/15 5:30 PM	12/28/15 2:30 PM	6	0 wks	0 wks	Project manager,Design	PM,DO,ME,F2,T1,T2
11			H	Assembling a hydraulic system	4 days	12/28/15 2:30 PM	1/4/16 9:30 AM	7,10	0 days	0 days	Hydraulic engineer, Fitter 1,Fitter 2,Team	HE,F1,F2,TL1
12			I	Assembling sensors and PLC	3 days	12/28/15 2:30 PM	12/31/15 10:30 AM	8,10	1.6 days	1.6 days	Electrical engineer, Fitter 5,Technician 3,Tec	EE,F5,T3,T4,TL2,F4
13	◆		J	Wiring (electrical wiring, sensors, field network...)	4 days	1/4/16 9:30 AM	1/7/16 3:00 PM	11,12	0 days	0 days	Electrical engineer, Computer engineer,Fitt	EE,CE,F4,F5,T1,T2
14			K	Developing and coding PLC application	4.24 wks	10/1/15 8:30 AM	10/28/15 4:00 PM	8	16.17 wks	16.17 wks	Electrical engineer, Developer 1	EE,D1
15			L	Developing and coding computer application	5.29 wks	10/1/15 8:30 AM	11/4/15 12:00 PM	8	14.64 wks	18.95 wks	Computer engineer, Developer 2	CE,D2
16	◆		M	Linking network to information system	1 day	1/7/16 3:00 PM	1/8/16 2:00 PM	13,15	22 days	22 days	Computer engineer, Fitter 1,Fitter 3,Technici	CE,F1,F3,T3,T4
17	◆		N	Integration mechanics, hydraulics, automation	16 days	1/7/16 3:00 PM	1/27/16 10:00 AM	12,13,14	0 days	0 days	Project manager, Mechanical engineer,Ele	PM,ME,EE,HE,CE,F1
18				◢ Tests and checks	166.47 day	9/24/15 9:30 AM	2/12/16 10:00 AM		0 days	0 days		
19			O	Network test	2.87 days	1/27/16 10:00 AM	1/29/16 2:00 PM	16,17	0 days	0 days	Project manager, Mechai	PM,ME,CE,T1,T2
20			P	Testing and checking prototype	2.29 wks	1/29/16 2:00 PM	2/12/16 10:00 AM	19,21	0 wks	0 wks	Project manager, Electrical engineer,Hydr	PM,EE,HE,CE,D1,D2
21	◆		Q	Technical documentation	2.24 wks	9/24/15 9:30 AM	10/8/15 9:30 AM	5	26.47 wks	26.47 wks	Technician 2,Technician	T2,T4,5E
22				◢ Marketing	11.2 days	2/12/16 10:00 AM	2/22/16 4:00 PM		0 days	0 days		
23			R	Finalizing technical documentation	2.27 days	2/12/16 10:00 AM	2/16/16 10:00 AM	20,21	5.2 days	5.2 days	Technician 4, Secretary-editor	T4,5E
24			S	Commercial documentation	1.13 wks	2/12/16 10:00 AM	2/19/16 10:00 AM	20,21	0 wks	0 wks	Marketing,Commercial r	MK,CS,QC
25			T	Presenting to client and approval	1.93 days	2/19/16 10:00 AM	2/22/16 4:00 PM	23,24	0 days	0 days	Project manager, Mechanical engineer,Ele	PM,ME,EE,HE,CE,CS
26				End of the project	0 days	2/22/16 4:00 PM	2/22/16 4:00 PM	25	0 days	0 days		

Figure 12.13. *The table with the columns "Resource names" and "Resource"*

In Figure 12.13, you can see that the task durations have been recalculated as well as the dates. The same goes for the free and total slack.

Figure 12.14. *The new slack calculated and the Gantt chart with the resource initials. For a color version of the figure, see www.iste.co.uk/reveillac/logistics.zip*

In the indicators column i on the left, you can see small logotypes in the shape of a character, which mark several rows (tasks G, J, M, N and Q). They are there to indicate overallocation of allocated resources. We will have to resolve this in order to keep a project coordinated while still trying to keep within the bounds of our constraints.

It is rare for overallocation not to be detected in a complex project. Microsoft Project provides several tools for overcoming these problems.

12.2.9. *Solving the problem of overallocation*

There are several possible solutions to resolve the problem posed by overallocation:

– replanning the task on the next available date for the resource;

– replacing this resource with another that may be available for the same work;

– acquiring a new resource to do this work (add a resource);

– moving the task in time by playing with the slack.

With the help of the team planner or the resource use table, we can see the periods of overallocation for each resource in detail. They are outlined in red, in front of each relevant resource. The time axis located along the top allows us to determine the dates and periods during which the normal work charge is exceeded ("attached" to the timetable that the resource should observe).

In this case, in Figure 12.15, we can see overallocation of the resources "Computer engineer", "Fitter 1", "Fitter 3" and "Technician 2".

In the resource use table, we can see these same overallocations in detail, with the actual number of hours work to be carried out.

By setting this display, we can obtain a great deal more information as shown in Figure 12.17 in which, for each resource and each task, the following elements appear: "Work", "Overallocation", "Cost" and "Remaining availability".

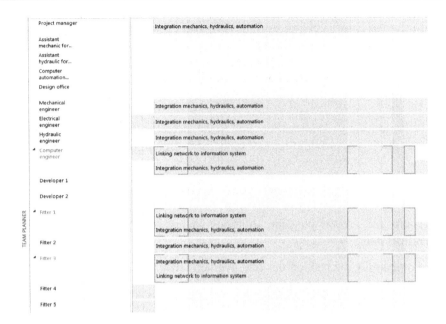

Figure 12.15. *The team planner. We can see the periods of overallocation for the resources "Computer engineer", "Fitter 1" and "Fitter 3". For a color version of the figure, see www.iste.co.uk/reveillac/logistics.zip*

		Presenting to client and approval	9 hrs	Work					
9	⊕	◢ Computer engineer	734.5 hrs	Work	7h	8.5h	8.5h	11h	13.5h
		Basic design	216 hrs	Work					
		Detailed design	119.5 hrs	Work					
		Automatic computer design	41.5 hrs	Work					
		Wiring (electrical wiring, sensors, field network…)	30 hrs	Work	7h	8.5h	8.5h	6h	
		Developing and coding computer application	187.5 hrs	Work					
		Linking network to information system	7.5 hrs	Work				2.5h	5h
		Integration mechanics, hydraulics, automation	60 hrs	Work				2.5h	8.5h
		Network test	17 hrs	Work					
		Testing and checking prototype	46 hrs	Work					
		Presenting to client and approval	9.5 hrs	Work					
10		◢ Developer 1	200.5 hrs	Work					
		Developing and coding PLC application	150 hrs	Work					
		Testing and checking prototype	50.5 hrs	Work					
11		◢ Developer 2	245.5 hrs	Work					
		Developing and coding computer application	187.5 hrs	Work					
		Testing and checking prototype	58 hrs	Work					
12	⊕	◢ Fitter 1	167.5 hrs	Work	1.5h			5h	13.5h
		Assembling a hydraulic system	30 hrs	Work	1.5h				
		Linking network to information system	7.5 hrs	Work				2.5h	5h
		Integration mechanics, hydraulics, automation	65 hrs	Work				2.5h	8.5h
		Testing and checking prototype	65 hrs	Work					
13		◢ Fitter 2	421 hrs	Work	1.5h			2.5h	8.5h
		Mechanical construction	321 hrs	Work					
		Assembling a hydraulic system	30 hrs	Work	1.5h				
		Integration mechanics, hydraulics, automation	70 hrs	Work				2.5h	8.5h
14	⊕	◢ Fitter 3	82.5 hrs	Work				5h	13.5h
		Linking network to information system	7.5 hrs	Work				2.5h	5h
		Integration mechanics, hydraulics, automation	75 hrs	Work				2.5h	8.5h
15		◢ Fitter 4	52.5 hrs	Work	7h	8.5h	8.5h	6h	
		Assembling sensors and PLC	22.5 hrs	Work					
		Wiring (electrical wiring, sensors, field network…)	30 hrs	Work	7h	8.5h	8.5h	6h	

Figure 12.16. *The resource use table with the number of hours for each resource. For a color version of the figure, see www.iste.co.uk/reveillac/logistics.zip*

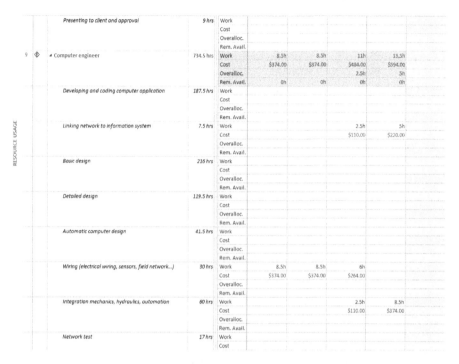

	Presenting to client and approval	9 hrs	Work					
			Cost					
			Overalloc.					
			Rem. Avail.					
9 ◇	◢ Computer engineer	734.5 hrs	Work	8.5h	8.5h	11h	13.5h	
			Cost	$374.00	$374.00	$484.00	$594.00	
			Overalloc.			2.5h	5h	
			Rem. Avail.	0h	0h	0h	0h	
	Developing and coding computer application	187.5 hrs	Work					
			Cost					
			Overalloc.					
			Rem. Avail.					
	Linking network to information system	7.5 hrs	Work			2.5h	5h	
			Cost			$110.00	$220.00	
			Overalloc.					
			Rem. Avail.					
	Basic design	216 hrs	Work					
			Cost					
			Overalloc.					
			Rem. Avail.					
	Detailed design	119.5 hrs	Work					
			Cost					
			Overalloc.					
			Rem. Avail.					
	Automatic computer design	41.5 hrs	Work					
			Cost					
			Overalloc.					
			Rem. Avail.					
	Wiring (electrical wiring, sensors, field network...)	30 hrs	Work	8.5h	8.5h	6h		
			Cost	$374.00	$374.00	$264.00		
			Overalloc.					
			Rem. Avail.					
	Integration mechanics, hydraulics, automation	60 hrs	Work			2.5h	8.5h	
			Cost			$110.00	$374.00	
			Overalloc.					
			Rem. Avail.					
	Network test	17 hrs	Work					
			Cost					

Figure 12.17. *A more detailed display for the resource use table. For a color version of the figure, see www.iste.co.uk/reveillac/logistics.zip*

As for our "Computer engineer", we will delete him in no. 13, "Wiring (electrical, sensors, field network, etc.)", since, after discussion with his coworkers, he states that the work can be carried out without his involvement.

After this deletion, the overallocation on task 13 disappears.

In task no. 21, technical documentation, we will replace "Technician 2" by "Technician 3" who can carry out the same work and who is available on this date.

In this way, the overallocation that also affected task no. 10, "Mechanical construction" disappears. In fact, "Technician 2" was also used at the same time and can therefore now be dedicated entirely to this task.

		Task Mode	Task	Name	D
1				Project launch	0
2			A	Defining specifications	5.
3				◢ **Design**	99
4			B	Basic design	7.
5			C	Detailed design	4.
6			D	Mechanical design	2.
7			E	Hydraulic design	2.
8			F	Automatic computer design	1.
9				◢ **Constructing prototype**	1:
10			G	Mechanical construction	1:
11			H	Assembling a hydraulic system	4
12			I	Assembling sensors and PLC	3
13			J	Wiring (electrical wiring, sensors, field network…)	4
14			K	Developing and coding PLC application	4.
15			L	Developing and coding computer application	5.
16	🧍		M	Linking network to information system	1
17	🧍		N	Integration mechanics, hydraulics, automation	1€
18				◢ **Tests and checks**	1€
19			O	Network test	2.
20			P	Testing and checking prototype	2.
21			Q	Technical documentation	2.
22				◢ **Marketing**	11
23			R	Finalizing technical documentation	2.
24			S	Commercial documentation	1.
25			T	Presenting to client and approval	1.
26				End of the project	0

Figure 12.18. *The overallocations in tasks no. 10, 13 and 21 have disappeared*

There now remain only two tasks affected by overallocation no. 16, "Network computer system" and no. 17, "Mechanical, hydraulic, computer integration".

The resources "Computer engineer", "Fitter 1" and "Fitter 3" are over used. To overcome this problem, we will ask the software to reschedule task no. 16 for the next possible date.

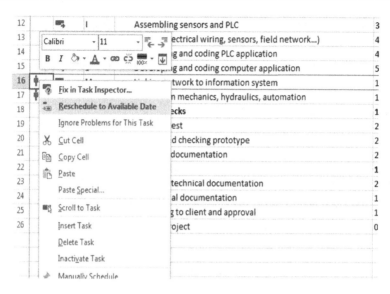

Figure 12.19. *Rescheduling 16 for the next possible date*

Once this operation has been carried out, everything is back in order and no task is affected by overallocation.

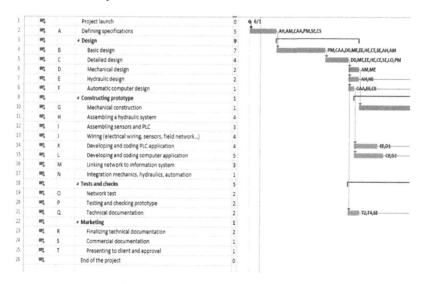

Figure 12.20. *No further task is overused. For a color version of the figure, see www.iste.co.uk/reveillac/logistics.zip*

We can now assume that the project has been planned correctly.

12.2.10. *Visualizing the project as a timeline*

In order to obtain a more general vision of the MecaTools project, it is possible to add either a simple or detailed timeline to the project.

For our example, we have chosen to put the summary tasks in legend form (above the time axis) and to display the other tasks in detail.

The critical tasks have been distinguished by a different background color.

Figure 12.21. *The detailed timeline for the "MecaTools" project. For a color version of the figure, see www.iste.co.uk/reveillac/logistics.zip*

12.2.11. *Using WBS coding*

Work breakdown structure (WBS) is a method of breaking down a project into a hierarchy with the tasks broken down into several levels. At present, it is often referred to as a project management flowchart.

It was the American Department of Defense that developed this concept in the 1950s. Later, it would be adopted by numerous businesses.

Its aim is to facilitate project organization.

In Microsoft Project, we can use a codification that depends on the principle of WBS. It can be customized. Its organization relies on an initial, main code which represents the project itself by default.

For our "MecaTools" project, we are going to create a code whose main prefix will be "1." followed by a series of ordered numbers.

With the help of the WBS tool in the ribbon, under the PROJECT tab, we will create the code structure.

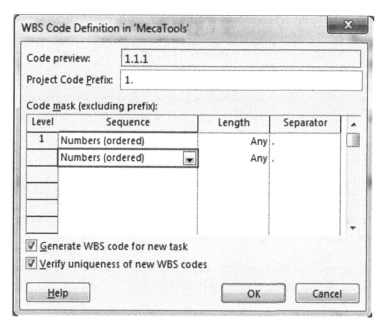

Figure 12.22. *The window for creating the WBS code structure*

We then add a WBS column in the table linked to the Gantt chart.

In Figure 12.23, we can see the hierarchy code that we have specified. It is implemented automatically in increasing order and by level depending on the tasks' importance (whether they are summaries or not).

	Task Mode	WBS	Task	Name	Duration	Start	Finish	Predecess
1		1.1		Project launch	0 days	6/1/15 8:0(	6/1/15 8:0(	
2		1.2	A	Defining specifications	5.88 wks	6/1/15 8:0(	7/7/15 8:3(	1
3		1.3		◢ Design	95.28 days	7/7/15 8:3(	10/7/15 8:(	
4		1.3.1	B	Basic design	6.05 wks	7/7/15 8:3(	8/21/15 5::	2
5		1.3.2	C	Detailed design	5.28 wks	8/21/15 5::	9/24/15 9:;	4
6		1.3.3	D	Mechanical design	2.23 wks	9/24/15 9::	10/7/15 8:(	5
7		1.3.4	E	Hydraulic design	2.23 wks	9/24/15 9::	10/7/15 8:(	5
8		1.3.5	F	Automatic computer design	1.12 wks	9/24/15 9::	9/30/15 2::	5
9		1.4		◢ Constructing prototype	108.74 day	9/30/15 2::	1/4/16 10:!	
10		1.4.1	G	Mechanical construction	10.7 wks	10/7/15 8:(	12/10/15 3	6
11		1.4.2	H	Assembling a hydraulic system	4 days	12/10/15 3	12/16/15 1	7,10
12		1.4.3	I	Assembling sensors and PLC	3 days	12/10/15 3	12/15/15 1	8,10
13		1.4.4	J	Wiring (electrical wiring, sensors, field network...)	4 days	12/16/15 1	12/21/15 4	11,12
14		1.4.5	K	Developing and coding PLC application	4.25 wks	9/30/15 2::	10/28/15 1	8
15		1.4.6	L	Developing and coding computer application	5.29 wks	9/30/15 2::	11/4/15 8:;	8
16		1.4.7	M	Linking network to information system	1.47 days	12/21/15 4	1/4/16 10:!	13,15
17		1.4.8	N	Integration mechanics, hydraulics, automation	8.8 days	12/21/15 4	1/4/16 8:2:	12,13,14
18		1.5		◢ Tests and checks	128.34 day	9/24/15 9::	1/13/16 9:!	
19		1.5.1	O	Network test	2.25 days	1/4/16 10:!	1/6/16 8:5(	16,17
20		1.5.2	P	Testing and checking prototype	1.27 wks	1/6/16 8:5(	1/13/16 9:!	19,21
21		1.5.3	Q	Technical documentation	2.27 wks	9/24/15 9::	10/8/15 9::	5
22		1.6		◢ Marketing	10.4 days	1/13/16 9:!	1/21/16 9:!	
23		1.6.1	R	Finalizing technical documentation	2.39 days	1/13/16 9:!	1/15/16 10	20,21
24		1.6.2	S	Commercial documentation	1.13 wks	1/13/16 9:!	1/20/16 9:!	20,21
25		1.6.3	T	Presenting to client and approval	1.27 wks	1/20/16 9:!	1/21/16 9:!	23,24
26		1.7		End of the project	0 days	1/21/16 9:!	1/21/16 9:!	25

Figure 12.23. *The task table with its WBS number column*

12.2.12. *Generating dashboards and reports*

There remains the possibility of editing the different dashboards to ensure monitoring and presentation of the project.

However, we can also generate reports to obtain an overview of resource use, the cost of tasks, etc.

Below, you will find some examples of possible dashboards and reports.

Microsoft Project offers numerous possibilities for creating different types of dashboards and graphs using raw data from the project or a pivot table of data that can be exported into a spreadsheet such as Microsoft Excel for future use.

12.2.12.1. Tasks/work time dashboard

This table is based on task use display It is possible to open a task to visualize the time details according to the resources, as shown in the "Mechanical construction" task in Figure 12.24.

Task Name	Work	Duration	Start	Details	', '15 T	S	Sep 21, '15 W	S	Oct 26, '15 T	
Project launch	0 hrs	0 days	l/15 8:0(	Work						
▷ Defining specifications	945.5 hrs	5.88 wks	l/15 8:0(	Work						
◢ Design	,240.5 hrs	95.28 days	'/15 8:3(	Work	719.25h	332.13h	179.13h			
▷ Basic design	,810.5 hrs	6.05 wks	7/15 8:3(	Work						
▷ Detailed design	,016.5 hrs	5.28 wks	1/15 5:2:	Work	719.25h	97.75h				
▷ Mechanical design	L50.25 hrs	2.23 wks	1/15 9:2:	Work		67.25h	83h			
▷ Hydraulic design	L50.25 hrs	2.23 wks	1/15 9:2:	Work		67.25h	83h			
▷ Automatic computer design	113 hrs	1.12 wks	1/15 9:2:	Work		99.88h	13.13h			
◢ Constructing prototype	,092.4 hrs	108.74 days	)/15 2:2:	Work			784.13h	935.5h	703h	3;
◢ Mechanical construction	2,188.5 hrs	10.7 wks	10/7/15	Work			408.63h	643.5h	694.63h	3;
Project manager	262.5 hrs		'/15 8:0£	Work			62.88h	99h	100.63h	
Design office	262.5 hrs		'/15 8:0£	Work			55.88h	88h	96h	:
Mechanical engineer	291.5 hrs		'/15 8:0£	Work			59.38h	93.5h	102h	:
Fitter 2	321 hrs		'/15 8:0£	Work			59.38h	93.5h	102h	(
Technician 1	350.5 hrs		'/15 8:0£	Work			59.38h	93.5h	102h	
Technician 2	372.25 hrs		'/15 8:0£	Work			59.38h	93.5h	102h	
Logistics	328.25 hrs		'/15 8:0£	Work			52.38h	82.5h	90h	
▷ Assembling a hydraulic system	120 hrs	4 days	)/15 3:5:	Work						
▷ Assembling sensors and PLC	135 hrs	3 days	)/15 3:5:	Work						
▷ Wiring (electrical wiring, sensors, fi	150 hrs	4 days	'15 10:5:	Work						
▷ Developing and coding PLC applicati	300.88 hrs	4.25 wks	)/15 2:2:	Work			187.75h	113.13h		
▷ Developing and coding computer ap	375 hrs	5.29 wks	)/15 2:2:	Work			187.75h	178.88h	8.38h	
▷ Linking network to information syst	26.03 hrs	1.47 days	1/15 4:2:	Work						
▷ Integration mechanics, hydraulics, a	797 hrs	8.8 days	1/15 4:2:	Work						
◢ Tests and checks	780.88 hrs	128.34 days	l/15 9:2:	Work		93.88h	151.5h			
▷ Network test	75 hrs	2.25 days	'16 10:5<	Work						
▷ Testing and checking prototype	460.5 hrs	1.27 wks	5/16 8:5<	Work						
▷ Technical documentation	245.38 hrs	2.27 wks	1/15 9:2:	Work		93.88h	151.5h			
◢ Marketing	199.5 hrs	10.4 days	1/16 9:5<	Work						
▷ Finalizing technical documentation	30.93 hrs	2.39 days	3/16 9:5<	Work						
▷ Commercial documentation	112.5 hrs	1.13 wks	3/16 9:5<	Work						
▷ Presenting to client and approval	56.08 hrs	1.27 days	)/16 9:5<	Work						
End of the project	0 hrs	0 days	l/16 9:5(	Work						

Figure 12.24. Table displaying the work time for each task

12.2.12.2. Resources/work time dashboard

This follows the same principle as the previous table, but is driven by the resources and it is possible to see details for each task, see the resource "Mechanical engineer" in Figure 12.25.

Resource Name	Work	Details	Aug 17, 15 M	W	F	Sep 21, 15 S	T	Oct 26, 15 T	S
▷ Unassigned	0 hrs	Work							
▷ Project manager	815 hrs	Work	24.5h	81.63h	65.88h		89.88h	108h	64
▷ Assistant mechanic for project	439.88 hrs	Work	63h	26.38h		70.63h	4.38h		
▷ Assistant hydraulic for project	439.88 hrs	Work	63h	26.38h		70.63h	4.38h		
▷ Computer automation assistant for project	337.5 hrs	Work	24.5h			37.5h			
▷ Design office	525 hrs	Work	38.5h	72.13h	40.38h		79.88h	96h	
⊿ Mechanical engineer	756.75 hrs	Work	48h	76.63h	35.88h	66.63h	93.5h	102h	7
Basic design	166.5 hrs	Work	48h						
Detailed design	112.5 hrs	Work		76.63h	35.88h				
Mechanical design	75.25 hrs	Work				66.63h	8.63h		
Mechanical construction	291.5 hrs	Work					84.88h	102h	7
Integration mechanics, hydraulics, automation	45 hrs	Work							
Network test	15 hrs	Work							
Testing and checking prototype	43.5 hrs	Work							
Presenting to client and approval	7.5 hrs	Work							
▷ Electrical engineer	637.5 hrs	Work	59.5h	81.63h	42.38h	66.13h	93.5h	27.88h	
▷ Hydraulic engineer	531.28 hrs	Work	59.5h	98.13h	42.88h	66.63h	8.63h		
▷ Computer engineer	680.93 hrs	Work	59.5h	101.5h	42.88h	66.63h	93.5h	65.38h	
▷ Developer 1	191.38 hrs	Work				27.13h	88h	35.75h	
▷ Developer 2	228 hrs	Work				27.13h	88h	72.38h	
▷ Fitter 1	137 hrs	Work							
▷ Fitter 2	412 hrs	Work					84.88h	102h	7
▷ Fitter 3	63.5 hrs	Work							
▷ Fitter 4	52.5 hrs	Work							
▷ Fitter 5	52.5 hrs	Work							
▷ Technician 1	500 hrs	Work					84.88h	102h	7
▷ Technician 2	563.38 hrs	Work				66.63h	93.5h	102h	7
▷ Technician 3	73.5 hrs	Work							
▷ Technician 4	130.13 hrs	Work				66.63h	18.5h		
▷ Secretary-editor	539.43 hrs	Work	52.5h	89.5h	51.88h	58.63h	16.5h		
▷ Marketing	37.5 hrs	Work							
▷ Commercial sector	240.5 hrs	Work							
▷ Logistics	548.75 hrs	Work		67.63h	75h	23.88h	74.88h	90h	6
▷ Quality control	150.53 hrs	Work							
▷ Team leader 1	91 hrs	Work							
▷ Team leader 2	83.5 hrs	Work							

Figure 12.25. *Table displaying the work time for each resource*

12.2.12.3. *Visual and chart reports*

The chart in Figure 12.26 has been created due to the report generator and compares the work for each of the resources with their availability.

To arrive at this result, it was necessary to extract data in order to then create a histogram, built by crossing multiple data (pivot table).

When the chart has been formatted, it is possible to save it in Microsoft Excel format, as an HTML page (in order to use it on a Website) or even as an ODS document (Open Document Spreadsheet, the document format for spreadsheets).

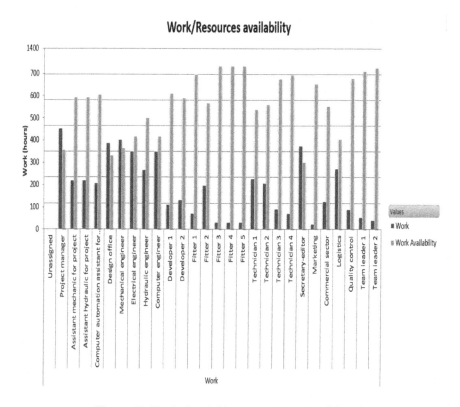

Figure 12.26. *A pivot table generated by the "Visual reports" tool from Microsoft Project 2013*

In Figure 12.27, you can see the Microsoft Excel spreadsheet that enabled the previous graph/chart to be created.

12.3. Monitoring the project

Once the project has commenced, the project leader can ensure that it is monitored. Each alteration, whether in tasks or resources is automatically taken into account and the planning is entirely recalculated (if the automatic planning mode is activated for the task or tasks in question).

As time passes and the project progresses along the calendar, we can select a task and give it an achievement percentage by means of the burndown tools available in the ribbon under the TASK tab, in the SCHEDULE group.

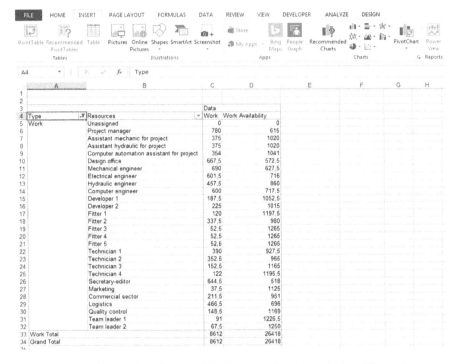

	A	B	C	D	E	F	G	H
1								
2								
3			Data					
4	Type	Resources	Work	Work Availability				
5	Work	Unassigned	0	0				
6		Project manager	780	615				
7		Assistant mechanic for project	375	1020				
8		Assistant hydraulic for project	375	1020				
9		Computer automation assistant for project	354	1041				
10		Design office	667,5	572,5				
11		Mechanical engineer	690	627,5				
12		Electrical engineer	601,5	716				
13		Hydraulic engineer	457,5	860				
14		Computer engineer	600	717,5				
15		Developer 1	187,5	1052,5				
16		Developer 2	225	1015				
17		Fitter 1	120	1197,5				
18		Fitter 2	337,5	980				
19		Fitter 3	52,5	1265				
20		Fitter 4	52,5	1265				
21		Fitter 5	52,5	1265				
22		Technician 1	390	927,5				
23		Technician 2	352,5	965				
24		Technician 3	152,5	1165				
25		Technician 4	122	1195,5				
26		Secretary-editor	644,5	518				
27		Marketing	37,5	1125				
28		Commercial sector	211,5	951				
29		Logistics	466,5	696				
30		Quality control	148,5	1169				
31		Team leader 1	91	1225,5				
32		Team leader 2	67,5	1250				
33	Work Total		8612	26418				
34	Grand Total		8612	26418				

Figure 12.27. *The Microsoft Excel spreadsheet, created by Microsoft Project, which enabled the chart in Figure 12.26 to be generated*

Figure 12.28. *The burndown and schedule tools available in the ribbon, under the TASK tab*

It is also possible to manage the project with an appropriate burndown by specifying a status date, PROJECT tab, STATUS DATE tool in the ribbon, and then update the project using the tool UPDATE PROJECT in the

STATUS group. In this case, Microsoft Project assumes that all the tasks have been fully completed at this point (by default).

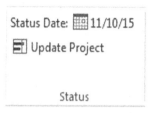

Figure 12.29. *The tools for managing the status date under the PROJECT tab*

The burndown state in creating tasks is visible on the Gantt chart, the task bars have a bold line down the center indicating the burndown.

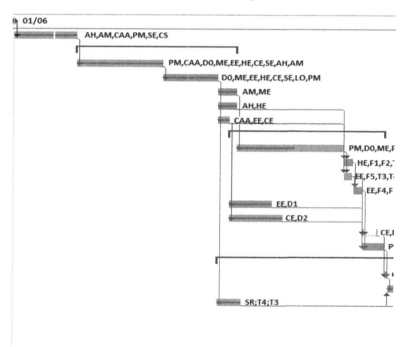

Figure 12.30. *Here, you can see the burndown (dark horizontal line in the center of the task bars). For a color version of the figure, see www.iste.co.uk/reveillac/logistics.zip*

To end, we can observe that the bars in the Gantt chart, which correspond to each task, are interactive, we can slide them to move the task to a different time (the cursor for this is in a crossed arrow shape), extend them to prolong the task duration (a cursor in the form of an arrow pointing right appears at the end of the bar) or even specify progress (a cursor in the shape of a % sign and a triangle appears at the start of the bar). Quite clearly, each change is noticed instantly and the project is recalculated as a consequence, if the automatic planning mode has been chosen for the task in question.

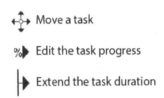

Figure 12.31. *The functions of the different types of cursors that appear when skimming over a bar representing a task in the Gantt chart*

12.4. To conclude

There could be a great deal more to say about the functions that Microsoft Project offers, however, we think that the items we have presented above show the main opportunities offered by this software.

We have deliberately omitted multiproject management, management using subprojects and resource sharing. To proceed further, to take advantage of these opportunities and to know more, go to the bibliography and the list of Internet links given at the end of this work.

13

Computerized Flow Simulation

13.1. To begin

Computerized flow simulation won acclaim with the expansion of logistics and as a result of the calculating power computers have now reached.

All types of flow can be modeled, whether they are discrete or continuous and many businesses use these simulation tools in lieu of and in place of systems or physical prototypes for which investments would be required.

Despite this, the software market hardly abounds with applications, of which there are few, but editors have garnered experience in the course of recent years and the functions they provide cover most domains (see section 9.4).

Historically, the first simulation software appeared in the 1990s (Scitor Process, ExtendSim, etc.). At this time, constraints on the machines' calculating power limited their potential and their ergonomics were far from convincing.

In this chapter, we have chosen intentionally to use an example that is not taken from an industrial setting, to show that logistics and flow simulation can be applied in the most unlikely domains.

13.2. The example we will use

To show you some of the opportunities offered by computerized flow management, especially with the software ExtendSim 9, we are going to create a simulation modeling the small winter sports resort, in France, "Levant" with its skiers, ski-lifts and pistes.

NOTE.– Our choice of the ExtendSim 9 software is due to the fact that we teach its use to our students in the department "Logistics and Transport Management" in the University of Burgundy –Faculty and Technology Institute Chalon sur Saône.

A version of this software, which will enable you to carry out the work, is available on the editor's Website (see Internet links at the end of the work).

We thank 1point2, the official distributor of "ExtendSim" software in France, for providing us with a US version.

13.2.1. *Resort map*

In Figure 13.1, you will find a map showing the whole area where skiing can take place, with its pistes and ski-lifts.

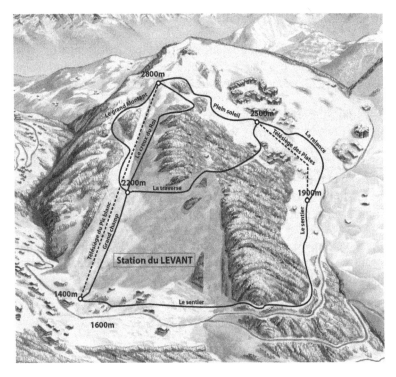

Figure 13.1. *The map of the "Levant" ski resort*

13.2.2. *Problem terms and specification*

The constraints to be added are as follows:

Resort opening hours: 9:00 am–5:30 pm.

Frequency with which skiers arrive at ski lifts, in seconds:

– 9:00 am–10:30 am: mean = 5 s, standard deviation: 1.5 s;

– 10:30 am–12:00 pm: mean = 14 s, standard deviation: 6 s;

– 12:30 am–1:45 pm: mean = 9 s, standard deviation: 4 s;

– 1:45 pm–2:30 pm: mean = 14 s, standard deviation: 6 s;

– 2:30 pm–5:30 pm: mean = 20 s, standard deviation: 7.5 s.

Ski-lift specifications:

– "Pic blanc" chair lift:

  - seat capacity: four skiers,

  - lift duration: 9 min,

  - seat distribution frequency: 18 s,

  - staff: one operator;

– "Les Plates" chair lift:

  - seat capacity: three skiers,

  - lift duration: 6 min,

  - seat distribution frequency: 15 s,

  - staff: one operator.

Workforce required:

– two operators (one per chair lift:) who work from 9 am to 5.30 pm;

– hourly cost with charges for one operator: 18.25 $/h.

Skier descent time per piste:

– "Grand champ": mean = 5 min, standard deviation = 2.5 min;

– "La traverse": mean = 9 min, standard deviation = 3.5 min;

– "Le creux du Roi: mean = 6 min, standard deviation = 2.5 min;

– "Le grand Montant" mean = 7 min, standard deviation = 2 min;

– "Plein soleil": mean = 6 min, standard deviation = 1.5 min;

– "Le sentier": mean = 11 min, standard deviation = 3 min;

– "La relance": mean = 6.5 min, standard deviation = 1.75 min.

The probabilities of a skier making any given choice are the following:

– when they arrive at the resort:

  - leave from the "Pic blanc" chair lift: 64%,

  - leave from "Les Plates" chair lift: 36%;

– when they arrive at the bottom of the pistes "Grand champ" and "Le sentier":

  - go to the start of the "Pic blanc" chair lift: 88%,

  - go to the parking lot (leaving the pistes): 12%;

– when they arrive at the bottom of the piste "La relance":

  - go to the piste "Le sentier": 60%,

  - go to the "Les plates" chair lift: 28%,

  - go to the parking lot: 12%;

– from the start of the "Pic blanc" chair lift:

  - go to the piste, "Le grand Montant": 35%,

  - go to the piste, "Le creux du Roi": 20%,

  - go to the piste, "Plein soleil" 45%;

– from the start of "Les Plates" chair lift:

  - go to the piste, "La traverse": 62%,

  - go to the piste, "La relance": 38%;

– at the bottom of the piste "Plein soleil":

  - go to the piste, "La traverse": 56%,

  - go to the piste, "La relance": 44%.

Various constraints and observations:

Of course, the pistes shown on the map are descending pistes, the skiers are not going to go up the slopes. The altitudes mentioned on the map indicate the directions of descent.

The skiers return to their two chair lifts on the way and arrive with their vehicle, a parking lot is available in front of each.

When a skier arrives at the bottom of the two pistes "Grand champ" and "Le sentier", they can choose to stop ski-ing and return to their vehicle in the "Pic blanc" parking lot. The same goes for the bottom of the piste "La relance", where the skier can return to "Les Plates" parking lot.

As for the "Les plates" chair lift, it is possible to go back to its point of departure, by using the "Pic blanc" chair lift and then by descending the two pistes "Plein soleil" and "La relance".

Ski passes are issued to the skiers in the form of a badge that they wear around their necks.

To access the chairlifts, the skiers have to validate their pass each time they use them, using a timesheet in which they insert their badge. Two time sheets are available at the "Pic blanc" chair lift and only one at "Les plates" chair lift. Each control system can be approached by a queue that should consist of no more than 25 skiers. It takes a checking time estimated at an average of 5 s for each skier and a standard deviation of 3 s.

NOTE.– All these data have been analyzed on the ground over several days. The constraints on the technology and workforce have been given by the service in charge of winter sports at the local authority for the "Levant".

This simulation does not take account of buying and selling passes to skiers. However, each skier should have one in their possession to use the pistes and the lifts.

The simulation should take place over 1 day, from 8:30 am to 6:00 pm.

Warning: the schedules are managed on a 24 h schedule (0:00:00 to 23:59:59 for 1 day)

13.3. Entering the project in the ExtendSim 9 software

We will now construct the simulation in the Extend Sim software, to do this we will go through a number of phases as shown in section 9.4.1.

13.3.1. *Defining the main parameters*

To begin, we will define a new model: menu FILE, NEW MODEL.

Once the model has been created, we will define the simulation time parameters: menu RUN, SIMULATION SETUP.

Our simulation is discreet since its flow will be made up of skiers (whole and countable).

The time will be managed using a calendar, we will, therefore, check USE CALENDAR DATES and choose 04/01/2015 for the START as determined by the specifications as well as the time for starting the simulation, which is 8:30:00.

By default, the time unit will be the minute (GLOBAL TIME UNITS) since it seems the best suited to the various activities in our future simulation.

The END TIME is fixed at 570 min, which is 570/60 = 9.5 h, since our simulation should take place over a day lasting from 8:30:00 to 18:00:00.

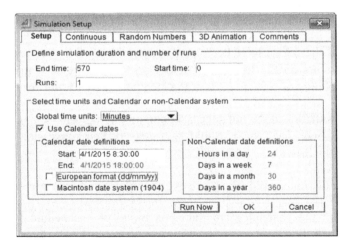

Figure 13.2. *Setting the parameters for our simulation*

To take account of our parameters, we will place the EXECUTIVE block from the "Item.lix" library in our model, in the corner above left.

Figure 13.3. *The new model with its EXECUTIVE block*

13.3.2. *Designing the model and entering the constraints*

With the help of the different blocks taken from the libraries (files with the extension ".lix"), we will now draw and construct the functional organization chart for our simulation. Incidentally, we will also use this to enter the constraints.

13.3.2.1. *Schedules and generators*

To simulate our skiers' arrival at the resort, we will use the CREATE block (library "Item.lix") whose function is to create entities in order to form a flow that, in our example, will be made up of skiers.

A generator can be linked to a statistical law and a schedule. In our specifications, we have four arrival frequencies for different skiers, managed by a normal law (mean and standard deviation), dependent on four schedules.

We should, therefore, place four CREATE blocks associated with four SHIFT blocks (library 'Item.lix") in our model.

Take, for example, the schedule from 9 am to 10:30 am, a skier arrives on average every 5 s with a standard deviation of 1.5 s, we will set the parameters for this block accordingly.

Let us begin by creating the schedule that we will call "morning schedule".

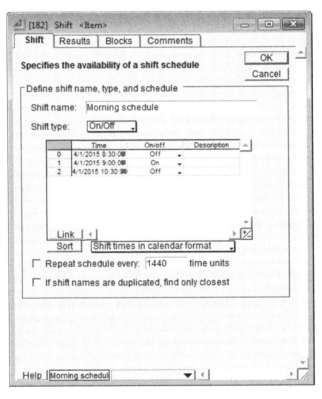

Figure 13.4. *The window for setting the parameters for the 9 am–10.30 am schedule*

In Figure 13.4, we can see the name, the schedule sections with a starting time of 8:30:00 – OFF, corresponding to the start time for our simulation which we have already entered in the SIMULATION SETUP window (section 13.3.1.), an arrival time for the skiers, 9:00:00 – ON and a time for switching to another frequency, 10:30:00 – OFF.

The schedule is in calendar format, which is easier to manage in our example since this format was chosen in SIMULATION SETUP.

NOTE.– A calendar should always have a start time equal to the start time entered in the simulation parameters.

To add or delete a row, you should simply click on the +/- box.

The ON position indicates the start of the active section of the schedule and the OFF position, its end.

The schedule can be in digital format, in this example according to the unit chosen in SIMULATION SETUP, we define the number of units of time passed since 0 (the moment the simulation commences).

On the same principle, we create three other CREATE blocks, associated with three other SHIFT blocks.

NOTE.– We can group the schedules 10:30–12:30 and 13:45–14:30 in a single SHIFT block since they have the same parameters for average and standard deviation.

We will also add another SHIFT block which will determine the operating times for the ski lifts and opening/closure of the pistes, which is 9:00 am–5:30 pm.

To summarize, the schedules and generators created are:

– "Morning schedule" with the generator "Morning arrival": 9:00–10.30;

– "Lunchtime schedule" with the generator "Lunchtime arrival": 12:30–13:45;

– "Day schedule" with the generator "Day arrival": 10:30–12:30 and 13:45–14:30;

– "End of day schedule" with the generator "end of day arrival": 14:30–17:30;

– "Resort schedule": 9:00–17:30.

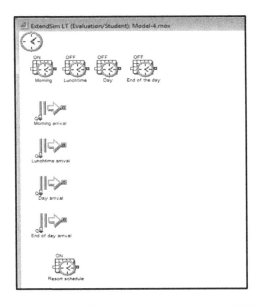

Figure 13.5. *The five SCHEDULE blocks and the four CREATE blocks*

13.3.2.2. *Checking passes*

Our skiers will now be directed toward one of the two lifts, the "Pic blanc" chair lift and "Les Plates" chair lift.

When they arrive at the chair lifts, they will have to validate their access with the timesheet and pass through the access turnstile.

Figure 13.6. *Two time sheets for checking passes with the access turnkey*

The "Pic blanc" chair lift has two time sheets and the "Les Plates" chair lift, one timesheet, which we should place before the queues (QUEUE block, library "Item.lix") in order to welcome the skiers.

In the specifications, we are told that these queues should not exceed 25 skiers. We cannot formalize this constraint, which is entirely dependent on our simulation and which we can instead consider as a result. We will discuss this again during the report and analysis phase (see section 13.3.7).

In order to bring the skiers to our chairlifts, we will put in place a SELECT ITEM IN (four entrances/one exit, library "Item.lix"), and then a SELECT ITEM OUT (we will name them: "Chair lift choice", library "Item.lix") in order to take account of the probabilities of skiers being redirected, which are 64% toward "Pic blanc" and 36% toward "Les plates".

Another SELECT ITEM OUT (we will name it: "Queue choice") will be placed in front of the queues (QUEUE block) at the "Pic blanc" in order to direct the skiers toward the shorter queue (logically, the skier will choose to go where there are fewer people).

Since "Les Plates" chair lift has only one queue, it should be enough to link the QUEUE + ACTIVITY blocks.

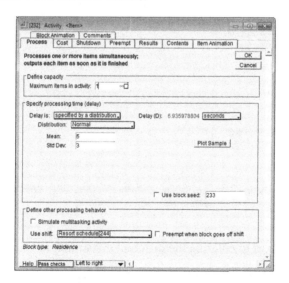

Figure 13.7. *Parameters for the ACTIVITY blocks "Pass checks" (distribution: normal, mean = 5, standard deviation= 3)*

Checks on passes are specified via the ACTIVITY blocks. The parameters for these blocks will be set according to normal distribution with an average of 5 s and a standard deviation of 3 s for one skier (as indicated in the specifications). The "resort schedule" defined previously will also be used.

Once all the blocks have been placed, we can link them to one another in order to obtain the model in Figure 13.8.

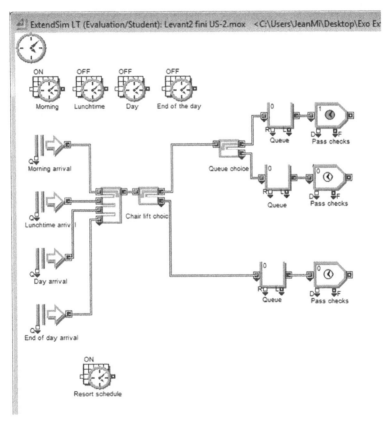

Figure 13.8. *Our model with the queues, pass checks, selectors and connections*

Let us now turn to entering the different constraints.

On the SELECT ITEM OUT block "Chair lift choice", which directs the skiers toward one or the other of the chair lifts, let us enter 0.64 (64%) toward "Pic blanc" and 0.36 (36%) toward "Les plates".

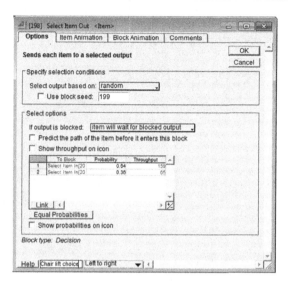

Figure 13.9. *The selector parameters "Chair lift choice" (probability: 0.64 and 0.36)*

For the "queue choice" selector, there arises the problem of how to direct the skier toward the shortest queue.

To resolve this difficulty, we will place a MAX & MIN block, from the library "Value.lix", in our model. Then, we will direct the value (the number of skiers), contained in each of the two queues toward it, so that it directs the skier to the most appropriate queue, (i.e. the shortest).

The parameters for the MAX & MIN block are the same as in Figure 13.10: OUTPUT THE: MINIMUM VALUE.

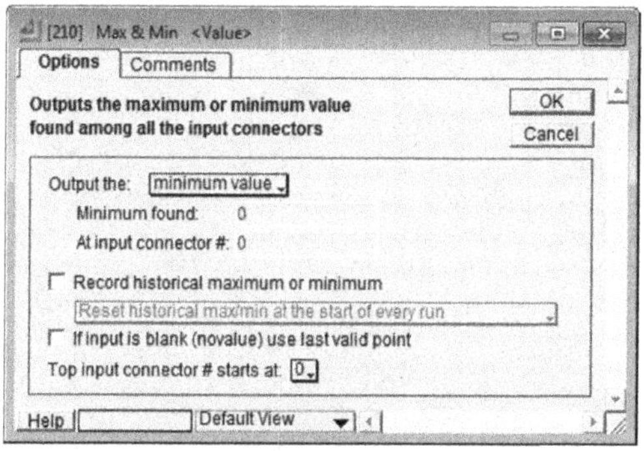

Figure 13.10. *The MAX & MIN block with its parameters set (minimum value)*

The parameters for the "Queue choice" selector are those in Figure 13.11: CHOOSE EXIT: BY SELECT CONNECTOR.

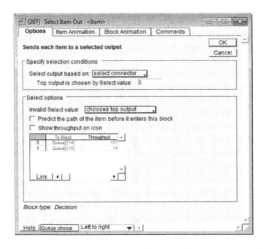

Figure 13.11. *The exit block selector "Queue choice" with parameters set (by SELECT CONNECTOR)*

Our model should resemble that in Figure 13.12.

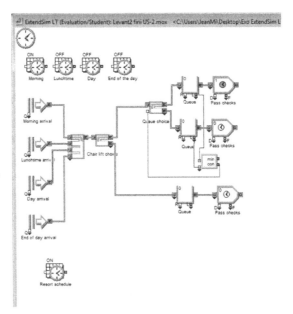

Figure 13.12. *Our model with the pass checks*

NOTE.– In order to simplify and reduce the model, we can use labels to replace the connections, for example in our model the labels Q1 and Q2 manage the queues and redirection of skiers for the "Pic blanc" chair lift.

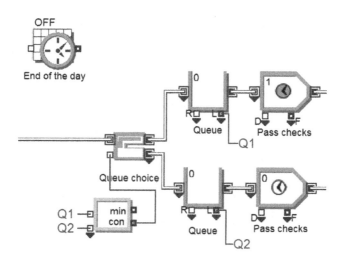

Figure 13.13. *The two labels Q1 and Q2 simulating the connections between the QUEUE blocks and the MIN & MAX block for checking passes for "The White Peak" chair lift*

13.3.2.3. *Transporting skiers*

Once their passes have been checked, the skiers wait in groups of four for "Pic blanc" chair lift and in groups of three for "Les Plates", for the operator to give them a seat. When they are seated, the chair lift will bring them to their destination, where each of them will choose a piste.

We are going to begin by making up batches of three or four skiers, representing a seat, via a BATCH block (library "Item.lix") whose parameters we will set by entering QUANTITY NEEDED.

Behind each of these BATCH blocks, we will place an activity that will simulate seat distribution to skiers. This will function according to the resort schedule and will have a constant duration of 18 s for "Pic blanc" chair lift and 15 s for "Les Plates" chair lift.

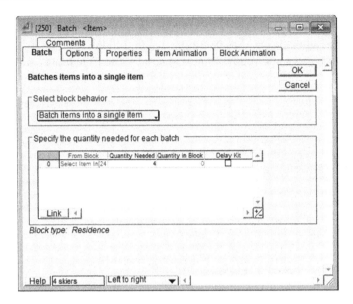

Figure 13.14. *Parameters for the BATCH block (quantity required = 4) for four skiers ("Pic blanc")*

Figure 13.15. *Parameters for the seat distribution blocks (schedule: "Resort schedule" and delay: constant at 18 s)*

The skiers are transported by the chair lift to the summit they have chosen, the duration of this transfer is 9 min for "Pic blanc" and 6 min for "Les Plates". For this action, we will use a CONVEY ITEM block (library "Item.lix").

These conveyors' capacity will be:

– for "Pic blanc": *(9 × 60s)/18s = 30* batches (of four skiers);

– for "Les Plates": *(6 × 60s)/15s = 24* batches (of three skiers).

The schedule used will be the "Resort schedule".

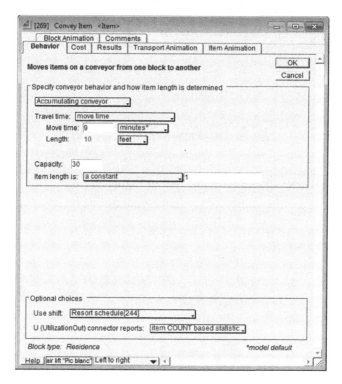

Figure 13.16. *Parameters for the CONVEY ITEM block simulating "Pic blanc" chair lift (transport time, duration = 9 min, capacity = 30, schedule: "Resort schedule")*

When the chair lift arrives, the skiers should be able to go their separate ways, we will, therefore, break up the batch (seat) using an UNBATCH block (library "Item.lix").

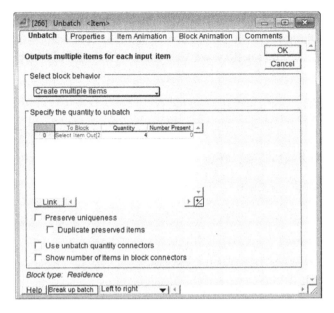

Figure 13.17. *Parameters for breaking down "Pic blanc" batch with an UNBATCH block (quantity = 4)*

Our new model should resemble that in Figure 13.18.

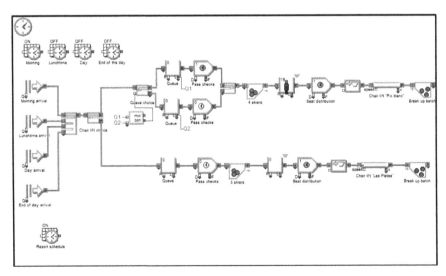

Figure 13.18. *Our model from the skiers' arrival until the chair lifts arrive at the top of the pistes*

13.3.2.4. *Ski pistes*

The skiers have now arrived at the top of the pistes and will be able to ski, we should, therefore, model all the pistes with their constraints: possible tracks, descent time and choice.

Each piste will be represented by an ACTIVITY block and a SELECT ITEM IN and SELECT ITEM OUT blocks will be used to formalize possible choices of direction (see Figure 13.1 the resort map).

Figure 13.20 depicts the part of the model that simulates the connection between different pistes. The SELECT ITEM IN blocks situated to the left will be connected to the two UNBATCH blocks, the one above to "Pic blanc" branch and the one below to "Les Plates" branch.

We will also include two EXIT blocks to depict return to the parking lot when the skiers decide to stop ski-ing.

Two SELECT ITEM OUT (library, "Item.lix") are placed in front of the exits, enabling the skier to continue skiing by returning to each of the two chair lifts.

The probabilities for each of the selectors are as indicated in the specifications.

All the pistes have the: "Resort schedule" opening times.

As the maximum number of skiers on the pistes has not been defined, we will assume that it is infinite.3

To redirect the skiers toward the chair lifts from the selectors situated behind the pistes: "Grand champ", "Le sentier" and "La relance", and behind the "Chair lift choice" selector, we will add two entrance selectors "Towards Pic Blanc" and "Towards Les Plates", as defined in Figure 13.21 and will link them all together.

13.3.3. *Defining flows*

The design and setting of parameters with the constraints imposed by the specifications is now finished, we will have to define the flows that will be conveyed by our model's two-dimensional (2D) animation.

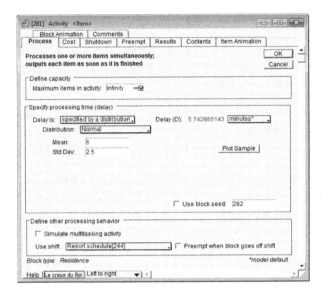

Figure 13.19. *An example of setting parameters for a piste, here "Le creux du Roi" (maximum number taking part in the activity: ∞, distribution: normal, mean = 6 min, standard deviation = 2.5 min)*

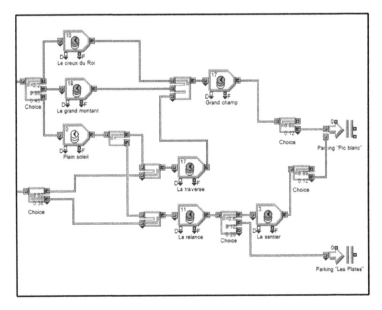

Figure 13.20. *The network of pistes, made up of the ACTIVITY, SELECT ITEM IN and SELECT ITEM OUT blocks. Note: the probabilities are mentioned on each of the EXIT SELECTOR blocks*

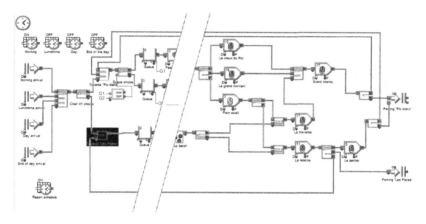

Figure 13.21. *The model with the redirection of skiers toward the chair lifts via the SELECT ITEM IN blocks: toward "Pic blanc" and "Les Plates"*

On each of the four generators, we will specify an entity type to animate, here we will choose PERSON_M. The flows circulating throughout our model will be made up of these entities, so we do not have to redirect them elsewhere, they spread automatically to all the blocks.

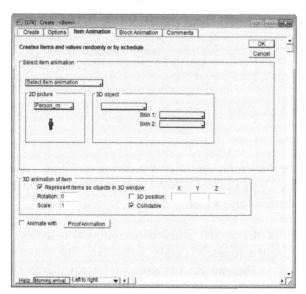

Figure 13.22. *Setting the flow parameters, here the entities transported are PERSON_M*

13.3.4. *Launching the simulation*

After checking 2D ANIMATION and ADD CONNECTION LINE ANIMATION in the SIMULATION menu, we can launch the simulation. To do this, you simply need to click on the launch icon in the tool bar above.

The simulation management functions are attached to certain icons whose functions are as follows:

– the green, triangular icon launches the simulation;

– the stop icon stops the simulation;

– the icon consisting of a double-bar + triangle pauses the simulation;

– the footprint icon continues the simulation step-by-step after it has been paused;

– the icon showing a green bouncing ball turns off the 2D animation while allowing the simulation to end;

– the rabbit-shaped icon accelerates the animation and reduces the simulation's total duration;

– the tortoise-shaped icon slows the animation and increases the simulation's total duration.

Figure 13.23. *The icons in the main tool bar that manage the simulation's progress*

NOTE.– To run the simulation in a shorter space of time, you can uncheck ADD CONNECTION LINE ANIMATION, the entities simulating the flows will only appear in some blocks.

Above left in the model window an egg-timer shows you the time remaining before the simulation ends, in minutes and seconds and the time taken compared to the real time entered in the parameters (here 570 min, from 8:30 am to 6:00 pm – the simulation duration determined by the specifications).

In some cases, the entities can evolve within the model. This often happens in industrial processes where components become subsets that can themselves become objects. Changes in entities, therefore, happen in some blocks and the items that leave are different from the items that enter.

13.3.5. Resource creation and allocation

Our simulation can function in its current state, however, the workforce, made up of the two operators, is still not included.

The role of these operators, allocated to each lift, is to allocate seats to the skiers when a line (a batch) of three or four skiers has been formed.

To overcome this constraint, we will have to create a RESOURCE POOL QUEUE in order to manage the workforce.

We will place this RESOURCE POOL QUEUE block (QUEUE block, library "Item.lix") in front of our ACTIVITY "Seat distribution".

The latter will be fed by a RESOURCE POOL block (library "Item.lix") made up of two operators that work during the resort's opening hours, from 9:00 am to 5:30 pm, at a cost, including charges of $18.25 per hour.

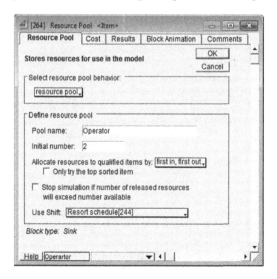

Figure 13.24. *RESOURCE POOL parameters "Operators" (Resource Pool name: Operator, Initial number: 2, schedule: "Resort schedule")*

Let us insert the RESOURCE POOL block in our model and enter the necessary parameters.

Figure 13.25. *Setting operator cost parameters (Cost = $18.25/h)*

When the RESOURCE POOL blocks have been created, we can insert the two blocks RESOURCE POOL QUEUE and allocate an operator to them.

Figure 13.26. *Parameters for the QUEUE RESOURCE blocks for the operators (RESOURCE FILE, Resource pool: Operator and Quantity = 1)*

Between each seat distribution, the operator should wait 15 or 18 s or even longer if there are no skiers present (which is unlikely). We will, therefore, remove them from operations during each waiting period, which is the role of the RESOURCE POOL RELEASE block (library "Item.lix").

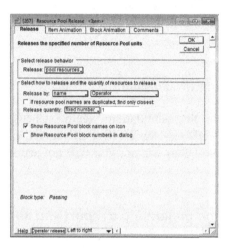

Figure 13.27. *Setting the parameters for the RESOURCE POOL RELEASE block (name: operator, fixed number = 1)*

We will place this block behind the ACTIVITY block "Seat distribution".

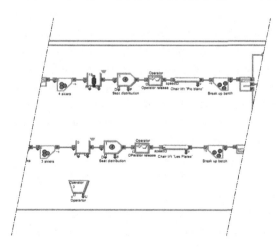

Figure 13.28. *The updated section responsible for managing chair lifts. We can see the RESOURCE POOL block (below), the two RESOURCE POOL QUEUE blocks and the two RESOURCE POOL RELEASE blocks*

Now that our model has been updated to include the two operators, it should resemble that in Figure 13.28.

NOTE.– The RESOURCE POOL QUEUE block is a QUEUE from the "Item.lix" library, whose parameters can be set to QUEUE RESOURCE mode. To mark this, the icon representing it is placed on a small green trapezium.

13.3.6. *Relaunching the simulation*

We can launch the simulation again. Check that SHOW 2D ANIMATION and ADD CONNECTION LINE ANIMATION are checked in the RUN menu, as if they are not you will not see any movement in the model.

13.3.7. *Creating and generating a report and analysis*

When our simulation model is operational, it is vital to have an overview of the data for the whole process in order to confirm that some resulting constraints have been met and can eventually bring about modifications.

This is the role of the report generator, which can synthesize all the results as a report in text form.

NOTE.– When a model is very complex, the simulation can take a very long time, in order to shorten this delay, we can ask ExtendSim to carry out the whole simulation without any animation being launched. In this case, the length of time the duration takes is much shortened, which has no effect on the calculations carried out and the values gathered by the software. To run a simulation without animation, you should simply uncheck 2D ANIMATION in the SIMULATION menu before launching.

To activate the report generator, you should check GENERATE REPORT in the RUN menu.

First, you can choose the type of report, of which there are three:

– Dialogues (Text): gives all the entrance-exit parameters for the blocks and the contents of the comment fields;

– Statistics (Text): gives all the final values for some exit parameters in the form of a table. It is designed to allow export to a spreadsheet;

– Statistics (BDD): the same as above, but it generates a database with tables. This database can be used and altered via the ExtendSim database management tool.

We are only interested here in the second type, "Statistics (text)", which will give us a coherent arrangement of the significant data gained from simulating our model.

To prevent our report from being overdetailed, we will choose, from the group of blocks, those figures which deserve our attention in terms of statistical results.

To begin, let us select the blocks "Arrival" (CREATE), "Queue" (QUEUE), "Pass checks" (ACTIVITY), "Chair lifts" (CONVEY ITEM), "Pistes" (ACTIVITY), "Operator" (RESOURCE POOL) and "parking lot" (EXIT). To do this, we hold down the CTRL key on the keyboard and we click on the each of the blocks.

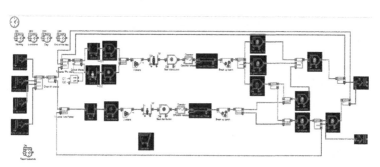

Figure 13.29. *Our model with the blocks selected for the report*

To attach these blocks to our future report, choose ADD SELECTED TO REPORT in the RUN menu.

Now, we launch the simulation and a dialogue box opens, asking for the report name ("report.txt" by default) and the file where we would like to save it.

By clicking on the SAVE button, the simulation will run, and then the report will display in the text window and save.

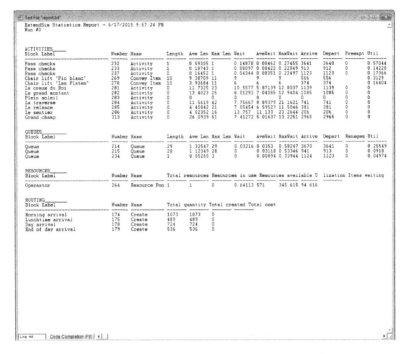

Figure 13.30. *The text window with our report in table form*

NOTE.– The GENERATE REPORT option in the RUN menu remains checked if it has already been activated and a new report is generated each time the simulation is launched.

A dialogue box will ask you its name. If you do not change it, the old report will be destroyed. It is, therefore, necessary to uncheck this option if you no longer want the report.

The form of the report will be kept if no change is requested.

Other options for the report are available in the RUN menu:"REMOVE ALL FROM REPORT" and "ADD ALL TO REPORT"…

In the specifications (section 13.2.2) and subsequently, when building the queues, (section 13.3.2.2), a queue size limited to 25 skiers has been requested.

Glancing at the report, we can see that the maximal length of three queues was 23, 14 and 22 skiers, respectively, which meets the constraint imposed.

13.3.8. *Development, enhancement and improvement*

We will continue by improving the process linked to our model. To do this, new constraints as well as some result indicators will be added.

13.3.8.1. *Skier categories*

Using figures taken over the previous season, the sports department for the Levant's local council has confirmed the following figures for us:

– 38% of skiers are men;

– 36% of skiers are women;

– 26% of skiers are children (under 16).

To divide the skiers into these three categories, (men, women and children), we will generate the categories randomly as the skiers arrive.

Let us begin by placing a RANDOM NUMBER block, taken from the "Value.lix" library in our model, beside the four "Arrival" ACTIVITY blocks as well as a SET block from the library "Item.lix".

We will link them as shown in Figure 13.31.

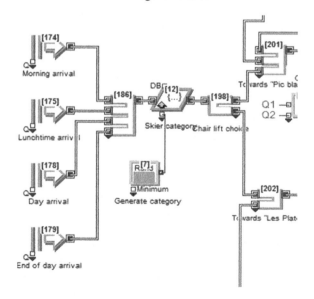

Figure 13.31. *In the center, the RANDOM NUMBER blocks ("Generate category") and SET ("Skier category") in our model*

NOTE.– In Figure 13.31, the no. of blocks appears, this is an option in ExtendSim, MODEL menu, check SHOW BLOCK NUMBERS.

After having opened the SET block, we will choose NEW STRING ATTRIBUTE in the PROPERTY NAME column in order to create our skier categories.

A dialogue box opens, in which we choose "CatSkier", which will become the general variable governing the three categories of skiers, and then we click on the OK button.

The dialogue box for the EXECUTIVE block opens, we see our variable in the table on the left, in the STRING ATTRIBUTE column.

Let us now enter our three categories: "Man", "Woman" and "Child" in the CATSKIER column in the right-hand table, as in Figure 13.32.

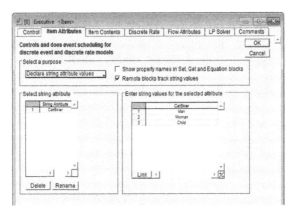

Figure 13.32. *The EXECUTIVE block dialogue box with its two tables containing the variable, string attribute "CatSkier" (table on the left) and the three categories, string values (table on the right)*

We should still allocate a different entity to each of our categories in order to make them differ visually on the 2D animation.

Let us open the SET block and select ITEM ANIMATION tab. After having chosen, CHANGE ITEM ANIMATION USING PROPERTY CatSkier from the two drop-down menus, we allocate the following entities on three rows (green +/- square), for each category:

– PROPERTY VALUE: "Man" 2D SYMBOL "Person_m";

– PROPERTYVALUE: "Woman", 2D SYMBOL: "Person_f ";

– PROPERTYVALUE: "Child", 2D SYMBOL: "Person_labor";

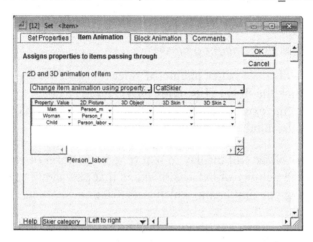

Figure 13.33. *Setting the parameters in the SET block for the categories (attributes) and their 2D symbols*

Then in the RANDOM NUMBER block, we will enter the percentages in the form of probabilities corresponding to the figures given, which are 0.38 (38%) for men, 0.36 (36%) for women and 0.26 (26%) for children.

The values are specified by distribution in an EMPIRICAL TABLE with three values (green +/- square).

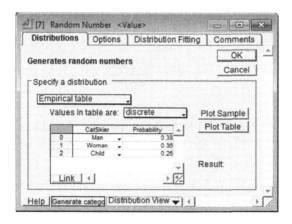

Figure 13.34. *Setting the parameters for the probabilities for our categories in the RANDOM NUMBER block*

We can launch a simulation after checking SHOW 2D ANIMATION and ADD CONNECTION LINE ANIMATION in the RUN menu. The various entities symbolizing the skier categories should appear on the connections and in some blocks.

13.3.8.2. *Chair lift monitoring curve*

In order to obtain a better picture of skier transport on each of the chair lifts, we will link the PLOTTER, DISCRETE EVENT block from the library "Plotter.lix" to the exit "u" of each of the CONVEY ITEM blocks. This exit "u" measures the number of entities conveyed.

This plotter block will display in real time in the form of two curves, the evolution of the number of batches of skiers that pass along each of the two chairlifts during our simulation, which lasts 570 min.

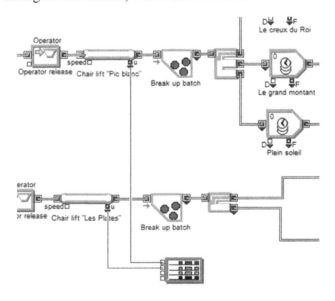

Figure 13.35. *The PLOTTER, DISCRETE EVENT block linked to the exits "u" of the CONVEY ITEM blocks (chairlifts)*

We must now set the parameters for this block.

When this block is opened, we find a tool bar in which we will be able to define the items necessary for setting the parameters.

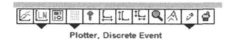

Plotter, Discrete Event

Figure 13.36. *The tool bar in the PLOTTER blocks*

Let us click on the icon on the far left (tool: "Trace properties"). A table with nine columns appears in which we can change, from left to right:

– the trace name describing the line;

– the line color (seven colors);

– the line width (five widths);

– the line pattern (four motifs);

– the line style (interpolated, stepped or plotted points);

– the line symbols (point, square, sign +, circle, etc.);

– the numerical format of the data table (general: Gen, decimal: x.xx, whole: xxx and scientific: x.xe);

– the axis of the line of the curve (Y1 on the left, or Y2 on the right);

– the show trace (visible: open eye, or invisible: closed eye).

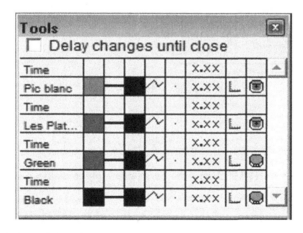

Figure 13.37. *The parameters to be entered and chosen in the line tool window. For a color version of the figure, see www.iste.co.uk/reveillac/logistics.zip*

Let us fill in the table as shown in Figure 13.37.

Let us now select the third tool from the left (tool: "Open dialog") and select its parameters by checking SHOW PLOT DURING SIMULATION (the curve line will create itself in real time during the simulation), SHOW INSTANTANEOUS QUEUE LENGTH…(the length of the queues will be visible instantaneously), AUTOSCALE DURING SIMULATION (the scale of the curve will be recalculated in real time as new data arrive) in the drop-down menu.

We can relaunch a simulation. The line window opens and the curve draws itself in real time through the simulation.

A table located below the curve shows the key figures (time and number of entities/batches) connected to each point on the curve for each chair lift.

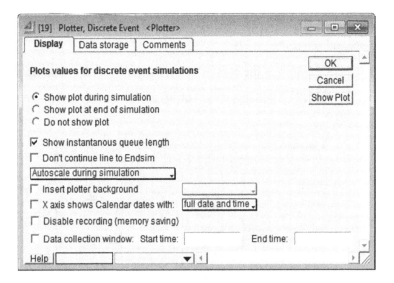

Figure 13.38. *Setting the parameters for the dialogue tool*

13.3.9. *Hierarchy*

ExtendSim software has an interesting feature called **hierarchy**. Due to this function, it is possible to regroup several blocks from the same model within a new block called **a hierarchical block** (or **H-block**).

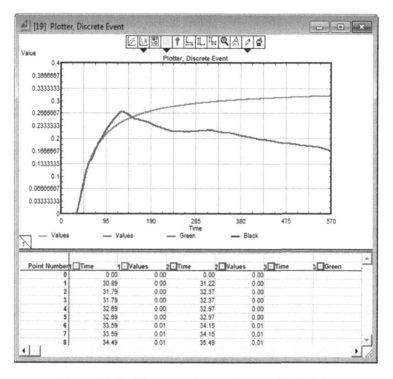

Figure 13.39. *The curve display window with its data monitoring table. For a color version of the figure, see www.iste.co.uk/reveillac/logistics.zip*

The appearance of this block can be changed to replace its default display (a simple gray square), in order to better suit its function.

In our model, we are going to create several **hierarchical** blocks:

– one block for the skiers' arrival and departure (parking lot);

– two blocks for the chair lifts;

– seven blocks for the pistes.

A **hierarchical** block uses all the entrances and exits already present in the model. In Figure 13.40, we can see the three entrances, attached to the pass checker for "Pic blanc", the three exits attached to the pistes and the exit for monitoring the number of skiers, attached to the PLOTTER, DISCRETE EVENT block.

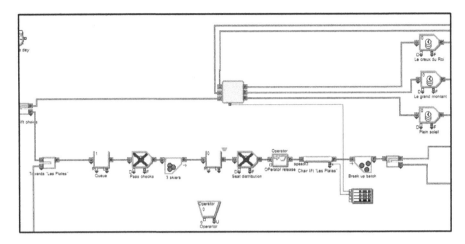

Figure 13.40. *The H-block (in the center) which is substituted for all the blocks for "Pic blanc" chair lift from the queues until the skiers' arrival on the pistes*

To build this block, we select the blocks to be broken down into hierarchies, and then we create an H-block via a right click, option MAKE HIERARCHICAL, which we name in the dialogue box that opens.

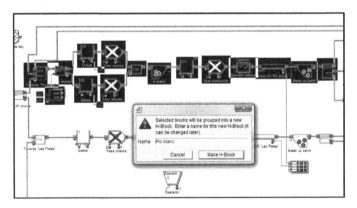

Figure 13.41. *Creating the H block for the "White peak"*

To open the block, we can choose OPEN STRUCTURE using a right click on the H-block. In the window (divided into four zones) that opens, we can add modifications, reorganize the entrances/exits, change the block's appearance and change the connections, etc.

We will not detail all the possibilities on offer here. They would far exceed the scope of this work.

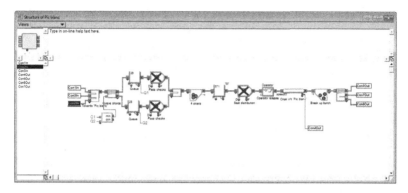

Figure 13.42. *The reorganized structure of "Pic blanc" H-block. You can see the graphical view (above left), the list of entrances/exits (below left), the help text (above right) and the submodel making up the H-block with its framed entrances/exits. For a color version of the figure, see www.iste.co.uk/reveillac/logistics.zip*

By applying the hierarchy technique, we can create the group of H blocks suggested above and modify their appearance in order to arrive at the model in Figure 13.43.

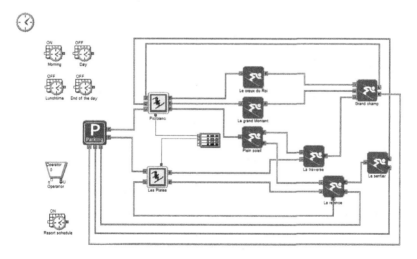

Figure 13.43. *Our model with all the H-blocks. The gray squares representing each of the blocks by default have all been replaced by logos*

13.3.10. *Design*

It is possible to improve the interface of a 2D model greatly by adding text, command buttons, menus, dialogue zones, etc., to it.

To arrive at this result, there are several tools:

– text, image and logo management;

– creating customized functions from the blocks in the library "Utilities.lix";

– cloning parameters or attributes from the dialogue boxes.

The upper tool bar has icons offering a whole range of graphic functions, some linked to particular menus.

Figure 13.44. *The bar tools that can be used, among other purposes, for design (from left to right): "Block/Text Layer", "Draw Layer", "Clone Layer", "All Layers", "Color", "Patterns", "Shapes", "Shuffle Graphics", "Cursor position" and "Icon tools"*

We will create:

– a title linked to a logo;

– a button to launch the simulation;

– a button to pause the simulation;

– two interactive dialogue zones to set the transport time parameters for each lift and the frequency of seat distribution.

Let us begin by entering the text for our title and our two dialogue zones (double-click on the bottom of the model and use the TEXT menu), as shown in Figure 13.45. The two frames are created using the "Draw Layer" tool and then colored with the "Color" tool. The logo is placed by copying/pasting a jpeg image that we will have chosen.

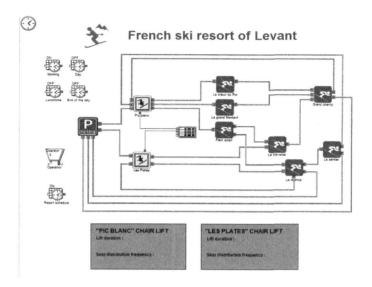

Figure 13.45. *The design items: title, logo, dialogue zones and frames*

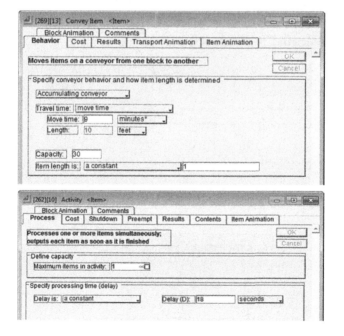

Figure 13.46. *The attributes "Move time … minutes*" to be cloned in the "chair lift" CONVEY ITEM (above) and "Delay (D)…seconds" to be cloned in the ACTIVITY blocks "Seat distribution" (below)*

In each of our zones we clone, using the "Clone Layer" tool from the toolbar, the attributes "Move time 9 minutes*" and "Move time 6 minutes*", the two "chairlift..." CONVEY ITEM blocks and the attributes "Delay (D) 18 seconds" and "Delay (D) 15 seconds" of the two ACTIVITY blocks "Seat distribution".

Finally, we should obtain dialogue zones identical to those in Figure 13.47.

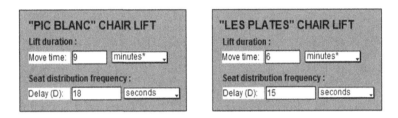

Figure 13.47. *The two finished dialogue zones*

To create our buttons, let us place two BUTTONS blocks taken from the library "Utilities.lix" in our model.

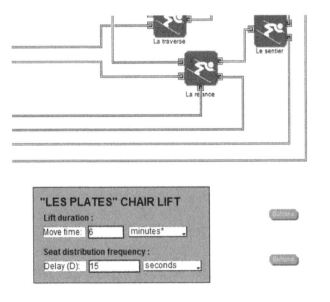

Figure 13.48. *Below right, two BUTTON blocks*

We will set the parameters for the first using RUN SIMULATION and for the second using PAUSE SIMULATION, and then we will clone (tool: "Clone Layer") each of the buttons in our model while taking care to place the clones on the BUTTON blocks to mask them.

Figure 13.49. *Setting the parameters for the "Run simulation" with the choice of button action and the cloning zone*

Once it is finished, our model should resemble that in Figure 13.50.

Our model is now finished and we can launch the simulation by clicking on the appropriate button, change the parameters for each of the chair lifts or even pause the simulation.

13.4. To conclude

ExtendSim presents numerous possibilities that are not shown here, such as the possibility of generating a three-dimensional (3D) animation using a model, creating its own blocks, manipulating databases via the add-in for Microsoft Excel, etc.

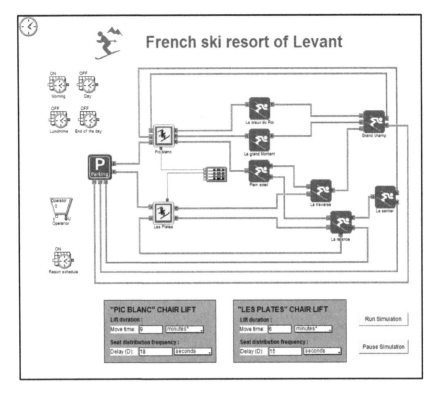

Figure 13.50. *Our model with its two buttons (below right)*
that mask the blocks

If you found the material in this chapter convincing and interesting, you can learn more by studying software assistance in more depth and by analyzing some of the examples provided.

A demonstration version of the ExtendSim software is downloadable from the editor's Website (see weblinks at the end of this book).

Other, equally powerful software exists, with some variations in how they are used, their functions and ergonomics (see section 9.4).

Conclusion

There is a fundamental question to be asked: what impact will operations research, flow management and the associated software tools have on the supply chains of tomorrow?

We can already list several impacts and the list is not exhaustive:

– innovation, by applying new practices which can be combined with implementing specialized software or specific modules within software that already exists to enhance its enterprise resource planning (ERP), for example;

– the acquisition of new methods and tools by staff in businesses, which will rely on an education programme necessary for it to be implemented completely;

– changes in procedures to further improve quality and to progress toward sustainable development;

– the implementation of more collaborative work so that actors in the logistics chain leave no room for uncertainty and so that data are handled from start to finish, whether within the business itself or with its subcontractors, partners and clients;

– restructuring of businesses' organizational operating methods to meet new criteria that arise from globalization and an openness to constant competition;

– the breakdown and overhaul of the supply chain according to the lifecycle of the product distributed, from its manufacture, via its storage, transport and distribution, and indeed its consumption, and its possible return or recycling;

– the linking of information flows, product flows and consumer behavior without, however, neglecting transparency and compliance and quality standards;

– the reduction in supply time without, however, compromising on flexibility in its processes.

The scope for taking advantage of operations research, decision-making aids and simulation has only grown in recent years, aided by technical progress in the optimization and power of information systems.

There has never been so much data to be handled. The explosion of systems and information technologies gave rise to data warehouses in the early 2000s to converge today toward "Big Data".

A short time ago, the user still had only a few constraints to handle in order to reach a unitary goal, today the number of variables to take into consideration has seen an exponential growth, many parameters that did not exist yesterday have now become indispensable, indeed fundamental. Faced with this situation, it has been necessary to refine choices, find relevant characteristics and effective handling methods.

Optimization tools and methods have become a necessity, indeed even a requirement for all actors in the supply chain.

In today's industrial world, a business's economic stability can fluctuate in a few days, indeed in a few hours, forcing rapid and strategic decision-making. Deciders have never had so much need for quantitative tools. The same goes for those in control of management and for a business's value-added services.

Logisticians are unanimous, in the years to come logistics, which is already a stake today, will have to progress still further. It will have to become simpler, more competitive and more secure.

The emergence, diversification and easy adaptability of simulation software to all industrial, commercial and logistical settings answer the problem of cost. It is no longer necessary to create a model or to implement an idea on the ground to check performance, everything is possible virtually for a modest investment and the variety of solutions is infinite.

In the next 30 years, there will be no revolution in the freight transport, trains, airplanes, boats and lorries that will still be here and it is a safe bet that there will be no new methods of transport.

However, it is realistic to imagine that each product will be fitted with intelligent microchips providing a more precise picture of its movement and use until it is destroyed. The internet of things is going in this direction and it is already a reality by means of technologies such as, among others, GPS[1] and RFID[2] chips. The information and data collected will bring new knowledge to the field which it will be necessary to handle, analyze, optimize and simulate to remain competitive.

Logistics has not stopped growing and investing in new, conceptual organizational territory, it will still increase in reliability and precision, this is its fate and because of this, it is also the fate of the industrial society of tomorrow. If, as we are writing this work, logistics seems to be everywhere, we can imagine that in the future, everything will be logistical.

---

1 Global positioning system, a global geolocation system invented in America, made up of a constellation of 24 satellites. It has been in use since 1995.

2 Radio frequency identification, a technology using radio-labels or transponders arranged around an electronic chip containing data readable at a short distance, as much as 10 m. A new generation has extended its reading zone to a radius of 200 m.

APPENDICES

Appendix 1

Installing the Solver

A1.1. Introduction

In Chapter 10, we used the Excel solver several times. By default, this is not always installed and does not, therefore, appear in the menus, the tool bar icons or the icons in the ribbon groups.

Below, you will find the procedures for installing it correctly for the different versions of Microsoft Excel for Microsoft Windows or Apple Mac OSX.

A1.2. Microsoft Excel 1997, 2002 and 2003 for Windows

Go to the TOOLS menu and choose ADD-INS

Figure A1.1. *The ADD-INS option...from the TOOLS menu*

A dialogue box opens, check "Add-Ins…"and click on the OK button.

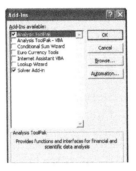

Figure A1.2. *The list of add-ins, with the "Solver Add-in" box to check*

To find the solver, select the TOOLS menu and choose SOLVER.

Figure A1.3. *The option for finding the SOLVER from the TOOLS menu*

A1.3. Microsoft Excel 2007–2010 for Windows

For Microsoft Excel 2007, click on the Office button (above right).

Figure A1.4. *The Office button in Excel 2007, top left*

In the window that opens, click on the EXCEL OPTIONS button (below right beside the QUIT EXCEL button).

For Microsoft Excel 2010, go to the FILE menu and choose OPTIONS.

In the dialogue box that opens, select "Adds-Ins" the list on the left.

Figure A1.5. *The Excel add-in with the drop-down MANAGE LIST (below, right) and the GO button*

In front of the "Manage" field (below), choose "Excel Add-Ins" in the drop-down list then click on the GO button.

In the "Excel add-ins" dialogue box that opens, check "Solver add-in" and click on the OK button.

Figure A1.6. *The dialogue box displaying the available add-ins*

The solver is the last available tool (on the right) in the ribbon under the DATA tab in the ANALYSIS group.

A1.4. Microsoft Excel 2013 for Windows

Go to the FILE menu and choose OPTIONS in the vertical ribbon situated on the left.

A dialogue box opens. In the list on the left, choose ADD-INS.

On the right, you should see all the Microsoft Office add-ins available, of which one is called "Solver Add-in".

Below, in front of the "Manage" field (below), select "Excel Add-ins" in the drop-down list then click on the GO button.

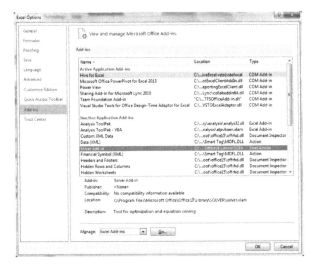

Figure A1.7. *The dialogue box showing the list of available add-ins on offer in Microsoft Excel 2013. Below, we can see the "Manage" field with its drop-down list and GO button*

A new dialogue box, called "Add-in", should open, in which a list is available.

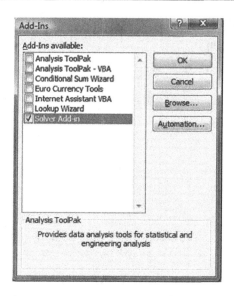

Figure A1.8. *The "add-ins" dialogue box*

Check "Solver add-in" and click on the OK button.

NOTE.– If "Solver add-in" does not appear in the list, click on the BROWSE button. From the file window that has opened, go to the list "Programmes\Microsoft Office\Office15\Library\SOLVER", select the file "SOLVER.XLAM" then click on the OK button.

To find the solver, select the DATA menu in the ribbon on the right in the ANALYSIS group, you should see the solver icon.

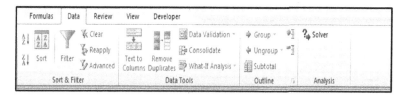

Figure A1.9. *The ribbon from the DATA menu, with the solver icon to the right*

A1.5. Microsoft Excel for Mac 2008–2011

Go to the TOOLS menu, choose ADD-INS.

In the "add-ins" dialogue box that appears, check "Solver.Xlam" and click on the OK button.

Figure A1.10. *The dialogue box for installing the solver add-in*

NOTE.– If "Solver.Xlam" does not appear in the "Add-ins available" list, click on the SELECT button, go to the "Applications" list, search for the folder "Microsoft Office 2008"or "Microsoft Office 2011", then the file "Office", then "Add-ins" and select "Solver.xla". Click on the OPEN button, it should appear in the list in the "Add-ins" dialogue box.

To find the solver from a spreadsheet, go to the TOOLS menu, and choose SOLVER...

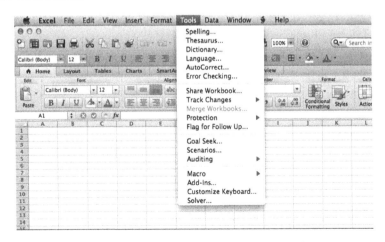

Figure A1.11. *The TOOLS menu and the SOLVER option... (at the very bottom)*

Appendix 2

Standard Normal Distribution Table

A2.1. Use

The following table carries the distribution function values of normal law. We read the whole number and the first decimal in the row, and the second decimal in the column.

For example:

For u = 0.92   F(u) = 0.8212

For u = -0.92   F(u) = 1 – 0.8212 = 0.1788

For F(u) = 0.908   u ≈ 1.33

For F(u) = 0.4362   1 – F(u) = 0.5638 and u ≈ 0.16

z	0.0	0.01	0.02	0.03	0.04	0.05	0.06	0.07	0.08	0.09
0,0	0.5000	0.5040	0.5080	0.5120	0.5160	0.5199	0.5239	0.5279	0.5319	0.5359
0.1	0.5398	0.5438	0.5478	0.517	0,5557	0.5596	0.5636	0.5675	0.5714	0.5753
0.2	0.5793	0.5832	0.5871	0.5910	0.5948	0.5987	0.6026	0.6064	0.6103	0.6141
0.3	0.6179	0.6217	0.6255	0.6293	0.6331	0.6368	0.6406	0.6443	0.6480	0.6517
0.4	0.6554	0.6591	0.6628	0.6664	0.6700	0.6736	0.6772	0.6808	0.6844	0.6879
0.5	0.6915	0.6950	0.6985	0.7019	0.7054	0.7088	0.7123	0.7157	0.7190	0.7224
0.6	0.7257	0.7291	0.7324	0.7357	0.7389	0.7422	0.7454	0.7486	0.7517	0.7549
0.7	0.7580	0.7611	0.7642	0.7673	0.7704	0.7734	0.7764	0.7794	0.7823	0.7852
0.8	0.7881	0.7910	0.7939	0.7967	0.7995	0.8023	0.8051	0.8078	0.8106	0.8133
0.9	0.8159	0.8186	0.8212	0.8238	0.8264	0.8289	0.8315	0.8340	0.8365	0.8389

z	0.00	0.01	0.02	0.03	0.04	0.05	0.06	0.07	0.08	0.09
1.0	0.8413	0.8438	0.8461	0.8485	0.8508	0.8531	0.8554	0.8577	0.8599	0.8621
1.1	0.8643	0.8665	0.8686	0.8708	0.8729	0.8749	0.8770	0.8790	0.8810	0.8830
1.2	0.8849	0.8869	0.8888	0.8907	0.8925	0.8944	0.8962	0.8980	0.8997	0.9015
1.3	0.9032	0.9049	0.9066	0.9082	0.9099	0.9115	0.9131	0.9147	0.9162	0.9177
1.4	0.9192	0.9207	0.9222	0.9236	0.9251	0.9265	0.9279	0.9292	0.9306	0.9319
1.5	0.9332	0.9345	0.9357	0.9370	0.9382	0.9394	0.9406	0.9418	0.9429	0.9441
1.6	0.9452	0.9463	0.9474	0.9484	0.9495	0.9505	0.9515	0.9525	0.9535	0.9545
1.7	0.9554	0.9564	0.9573	0.9582	0.9591	0.9599	0.9608	0.9616	0.9625	0.9633
1.8	0.9641	0.9649	0.9656	0.9664	0.9671	0.9678	0.9686	0.9693	0.9699	0.9706
1.9	0.9713	0.9719	0.9726	0.9732	0.9738	0.9744	0.9750	0.9756	0.9761	0.9767
2.0	0.9772	0.9778	0.9783	0.9788	0.9793	0.9798	0.9803	0.9808	0.9812	0.9817
2.1	0.9821	0.9826	0.9830	0.9834	0.9838	0.9842	0.9846	0.9850	0.9854	0.9857
2.2	0.9861	0.9864	0.9868	0.9871	0.9875	0.9878	0.9881	0.9884	0.9887	0.9890
2.3	0.9893	0.9896	0.9898	0.9901	0.9904	0.9906	0.9909	0.9911	0.9913	0.9916
2.4	0.9918	0.9920	0.9922	0.9925	0.9927	0.9929	0.9931	0.9932	0.9934	0.9936
2.5	0.9938	0.9940	0.9941	0.9943	0.9945	0.9946	0.9948	0.9949	0.9951	0.9952
2.6	0.9953	0.9955	0.9956	0.9957	0.9959	0.9960	0.9961	0.9962	0.9963	0.9964
2.7	0.9965	0.9966	0.9967	0.9968	0.9969	0.9970	0.9971	0.9972	0.9973	0.9974
2.8	0.9974	0.9975	0.9976	0.9977	0.9977	0.9978	0.9979	0.9979	0.9980	0.9981
2.9	0.9981	0.9982	0.9982	0.9983	0.9984	0.9984	0.9985	0.9985	0.9986	0.9986
3.0	0.9986	0.9987	0.9987	0.9988	0.9988	0.9989	0.9989	0.9989	0.9990	0.9990
3.1	0.9990	0.9991	0.9991	0.9991	0.9992	0.9992	0.9992	0.9992	0.9993	0.9993
3.2	0.9993	0.9993	0.9994	0.9994	0.9994	0.9994	0.9994	0.9995	0.9995	0.9995
3.3	0.9995	0.9995	0.9995	0.9996	0.9996	0.9996	0.9996	0.9996	0.9996	0.9997
3.4	0.9997	0.9997	0.9997	0.9997	0.9997	0.9997	0.9997	0.9997	0.9997	0.9998
3.5	0.9998	0.9998	0.9998	0.9998	0.9998	0.9998	0.9998	0.9998	0.9998	0.9998
3.6	0.9998	0.9998	0.9999	0.9999	0.9999	0.9999	0.9999	0.9999	0.9999	0.9999
3.7	0.9999	0.9999	0.9999	0.9999	0.9999	0.9999	0.9999	0.9999	0.9999	0.9999
3.8	0.9999	0.9999	0.9999	0.9999	0.9999	0.9999	0.999	0.9999	0.9999	0.9999
3.9	1.0000	1.0000	1.0000	1.0000	1.0000	1.0000	1.0000	1.0000	1.0000	1.0000

Table A2.1. *Standard normal distribution*

Glossary

Analytical resolution: in this type of problem solving, the solution is approached using a mathematical method that provides precise and rapid results. A graphic solution can be chosen in parallel with analytical resolution. In linear programming, the simplex method is a type of analytical resolution.

CAPM (assisted production computer management): software designed for managing the group of activities associated with production in a business. There, we can find, in the form of modules: order management, item management, resource management, name and line management, manufacture planning management, buying management, etc.

Canonical form: form of a linear programming problem in which the constraints are expressed by in equations.

Cardinality: a pair of values attached to a relation in the entity-relation model (relational model). For a relation (or association) between two entities, two pairs should be specified. The cardinalities express the number of times that an entity's occurrence should be taken into account in the relation, as a minimum or as a maximum. There are four typical pairs of cardinalities: *(0. 1)*, *(1. 1)*, *(0. n)* and *(1. n)*, *n* being equivalent to several.

Constraint: factor or condition that an optimization problem should satisfy.

CPM (critical path method): method of calculating a critical path in a graph for planning tasks.

Decision-making aid: group of techniques relying on probability theory, graph theory or even operations research. They offer a human actor the opportunity to opt for the best solution possible faced with largely industrial, financial or other problems.

Deterministic algorithm: algorithm that carries out a predetermined process to resolve a problem.

Dual: in linear programming the initial problem, called "primal", is linked to another linear problem known as "dual". "Dual" is in a way is symmetric to "primal".

Functions:

Affine function: this is a function of the form $y = ax + b$. The variable a is called the leading coefficient and the variable b, the intercept. An affine function is a line. If it passes the starting point, this is an exception (0 intercept) and so a linear function is obtained.

Economic function: function to be optimized in the context of linear programming.

Linear function: see Affine function.

Objective function: see Objective.

Heuristic: this describes a method that progresses via successive approaches to eliminate the alternatives so that only a narrow group of solutions, close to an optimum solution, is retained.

Laws:

Kirchhoff's law: also known as "node law". This is a law that shows the energy conservation in an electrical circuit. The sum of current intensities that enter a node is equal to the sum of current intensities that leave it. By extension, this law has a place in operations research, among other purposes, in flow management.

Normal law: also known as "Gauss's law" or "Laplace-Gauss" law. It is the law of probability most used in statistics. Its curve is called a "bell curve" on account of its shape. It has two parameters: an expected value and a real number that enables us to calculate the probabilities of a group of continuous random variables.

Maximization: this describes an optimization problem when its objective should tend toward a maximum.

Minimization: this describes an optimization problem when its objective should tend toward a minimum.

MPM (meter potential method): method invented by Bernard Roy in 1958. It enables us to illustrate and optimize the tasks in a project in the form of a network diagram showing, among other aspects, the interdependency between tasks while reducing delays and constructing a critical path. In MPM, the tasks are represented by nodes (vertices) in contrast to the PERT method.

MST (minimal spanning tree): a tree that connects all the nodes in a graph and in which the sum of the weights of each of the edges is minimal.

MWST: minimum weight spanning tree: see MST.

Objective: this is a shortened term for the objective function. This function is the main criterion for determining a solution in a mathematical optimization problem.

ODS (Open Document Spreadsheet): a file extension and open document format designed for databases, based on the ODF norm (Open Document Format for Office Applications). This format is used, in particular, by "Calc", the spreadsheet from OpenOffice suite.

OOP (Object Oriented Programming): computer programming method created in the1960s by O.J. Dahl and K. Nygaard, then further developed by A. Kay. It is built around objects, which are software modules that interact with one another and represent an idea, a concept or an entity existing in the physical world. A number of languages, such as Java, C++, PHP, Ada, etc., use this programming method. The idea of an object is also present within SEW (Software Engineering Workshops–Microsoft Visual Studio, Netbeans, etc.), in some modeling tools (UML, DBDesigner, etc.) and in distributed object technology (Corba, Pyro, RMI, etc.).

PEP (program evaluation procedure): planning method used by the American air force.

PERT (program evaluation and review technique or program evaluation research task): project planning method refined by Booz, Allen and Hamilton, at the end of the 1950s, at the request of the US Navy Special Projects Office. It enables a connected, directed and valued graph to be created, in which each task is represented by an edge that joins two vertices. It optimizes the project while still taking account of the functional constraints in order to determine the critical path and tasks.

Pivot: also called a "Gauss Pivot". The pivot is linked to the method of the same name (also called "Gauss-Jordan Elimination"). It is an algorithm that determines the solutions of a system of linear equations.

Primal: this describes the initial problem in linear programming.

RDBMS (relational database management systems): database management system which manages the relations forming the constraints, which guarantee referential data integrity.

Relation: this is equivalent to a link between two entities in a relational model. The relationship is established between two tables in an RDBMS.

Relational algebra: this is the mathematical and logical foundation on which the relational model is built. It suggests a group of elementary operations to create new relations.

Retro-active planning: also called, "Retroplanning" (*Feedback scheduling*). It is a reversed method of planning that imagines a project starting from the end date to then work back to the start. This method can be used when a project end date is fixed in advance and is imperative.

SEP (separation and evaluation procedure): also known as "branch and bound". A method that consists of seeking the optimal solution for a combinatorial optimization problem. It is built around the idea of separation (*branch*) which breaks down the group of solutions into smaller subgroups and around optimistic assessment to increase (*bound*) these same subgroups.

VBA (Visual Basic Application): the implementation of the Microsoft Visual Basic language within Microsoft Office applications. Some applications such as WordPerfect, SolidWorks or AutoCAD also include some of this language.

VBE (Visual Basic Editor): tool for editing and refining VBA language. This can also be the acronym for Visual Basic for Excel.

WBS (work breakdown structure): system of creating a hierarchical breakdown of the tasks in a project.

Bibliography

[BOL 02] BOLLOBAS B., *Modern Graph Theory,* Springer, 2002.

[BON 13] BONNIE B., *Microsoft Project 2013: The Missing Manual,* 1st ed., O'Reilly Media, 2013.

[COH 95] COHEN V., *La recherche opérationnelle,* Collection "Que sais-je?", Presses Universitaires de France, 1995.

[CRO 68] CROLAIS M., *Gestion intégrée de la production et ordonnancement,* Dunod, 1968.

[FAU 96] FAURE R., *Précis de recherche opérationnelle,* 3rd ed., Dunod, 1996.

[FRE 15] FRÉDÉRIC L.G., *Macros et langage VBA – Apprendre à programmer sous Excel,* 3rd ed., ENI, 2015.

[GAR 05] GARDARIN G., *Base de données,* 6th ed., Eyrolles, 2005.

[GON 09] GONDRAN M., MINOUX M., *Graphes et algorithmes,* 4th ed., Lavoisier, 2009.

[GOT 04] GOTHA G., *Modèles et algorithmes d'ordonnancement,* Ellipses, 2004.

[GRO 05] GROSS J.L., YELLEN J., *Graph Theory and Its Applications,* Chapman & Hall, 2005.

[HEN 10] HENRI L., *VBA Excel 2010 – Créez des applications professionnelles: Exercices et corrigés,* ENI, 2010.

[JAC 11] JACOB R.S., WILLIAM B., CLAY W. *et al., Manufacturing Planning and Control for Supply Chain Management,* McGraw-Hill, 2011.

[JAM 00] JAMES S., *Strategic Logistics Management,* 4th ed., McGraw-Hill, 2000.

[KOR 10] KORTE B., VYGEN J., *Optimisation combinatoire: Théorie et algorithmes,* Springer, 2010.

[MAR 13] MARIAPPAN P., *Operations Research: an Introduction*, Pearson, 2013.

[MIC 08] MICHELLE C., *Algèbre relationnelle: Guide pratique de conception d'une base de données relationnelle normalisée*, ENI EDS, 2008.

[MIC 14] MICHAEL P.L., *Scheduling: Theory, Algorithms, and Systems*, 4th ed., Springer, 2014.

[MOS 10] MOSHE S., *Dynamic programming: Foundations and principes*, 2nd ed., CRC Press, 2010.

[PHI 08] PHILIPPE V., *Problématique de la logistique*, Economica, 2008.

[PIE 91] PIERRE D.P., *Plein flux sur l'entreprise: La nouvelle logistique, de la gestion des stocks à la gestion des flux*, Nathan Entreprise, 1991.

[PRI 11] PRINS C., SEVAUX M., *Programmation linéaire avec Excel*, Eyrolles, 2011.

[RIC 97] RICHARD B., *Schaum's Outline of Operations Research*, 2nd ed., McGraw-Hill, 1997.

[VAJ 13] VAJDA S., *Linear Programming: Algorithms and Applications*, Springer, 2013.

[VED 85] VÉDRINE J.P., *T.Q.G.*, Vuibert, 1985.

[VIC 96] VINCENT G., *Gestion de la production*, 2nd ed., Economica, 1996.

[WAL 13] WALKENBACH, *Excel 2013 Power Programming with VBA*, 1st ed., Wiley, 2013.

[WAY 97] WAYNE L., WINSTON, *Operations Research Applications and Algorithms*, 3rd ed., Duxbury Pr, 1997.

Internet links

Internet links are transitory by nature. In the course of time they can move to other addresses, or even disappear. All were valid when this work was written, and if some of them no longer work, a short search on Google will help you to find them.

Graph theory and operations research

ELSEVIER: European Journal for Operations Research, http://www.journals.elsevier.com/european-journal-of-operational-research.

INSA ROUEN: Graph theory and operations research, https://moodle.insa-rouen.fr/course/view.php?id=124.

ROADEF: French Society for Operations Research and Decision-making aids, http://www.roadef.org/content/road/road.htm.

University of Texas: Operations research models http://www.me.utexas.edu/~jensen/models/index.html.

University of Nancy: Maximum flow in a graph, J.F. SCHEID, http://www.iecn.u-nancy.fr/~scheid/Enseignement/flotmax.pdf.

IEOR Berkeley: Network flows and graphs, http://www.ieor.berkeley.edu/~ieor266/Lecture14.pdf.

University of Metz: Modeling maximum flow problems, http://ensrotice.sciences.univ-metz.fr/module_avance_thg_voo6/co/modelflotmax.html.

ENSTA: Introduction to discrete optimization, Adam OUOROU, http://wwwdfr.ensta.fr/Cours/docs/MAE41/maxflow_hd.pdf.

Courses on graphs, paths, trees and flows, http://idmme06.inpg.fr/~rapinec/Graphe/Graphe/default.html.

Books on line: Applied Mathematical Programming, BRADLEY, HAX, and MAGNANTI (Addison-Wesley, 1977), http://web.mit.edu/15.053/www/.

Article: On the history of transportation and maximum flow problems, Alexander SCHRIJVER, http://homepages.cwi.nl/~lex/files/histtrpclean.pdf.

Introduction on graph theory, Didier MÜLLER, http://www.apprendre-en-ligne.net/graphes/.

CNRS: Graph theory and graph optimization, Christine SOLNON, http://liris.cnrs.fr/csolnon/polyGraphes.pdf.

INPL: Elements of graph theory and linear programming, Didier MAQUIN, http://cours.ensem.inpl-nancy.fr/cours-dm/graphes/Graphesnew.pdf.

University of Bordeaux: Introduction to graph, Bruno COURCELLE, http://www.labri.fr/perso/courcell/Conferences/GraphesX.pdf.

Graph theory, http://theoriedesgraphes.com/.

Optimization and complexity: A group of comprehensive courses, HUET, http://doc.ium.bz/S6/Optimisation%20et%20complexit%C3%A9/cour/.

Databases and relational algebra

University of Paris 8: Databases, Relational algebra, Rim CHAABANE, http://www.ai.univ-paris8.fr/~lysop/bd/seance5-ModeleRel-suite.pdf.

ENST: The relational model and relational algebra:, http://www.enst.dz/Cours/ LeModele_relationneletAlgebre_relationnelle.pdf.

Introduction to SQL and Alexandre MESLE language, http://alexandre-mesle.com/ enseignement/sql/index.html.

What is an RDBMS? Fabien CELAIA, http://fadace.developpez.com/sql/coddsgbdr/.

Tables, Pivot tables and VBA with Microsoft Excel

Excel-pratique, free VBA course, http://www.excel-pratique.com/fr/vba.php.

Documentation for developers VBA, Microsoft, https://msdn.microsoft.com/fr-fr/office/ff688774.aspx.

Excel exercise – The relations between tables, http://www.excel-exercice.com/excel2013/relation-entre-les-tables/.

How to create a table with Excel, WikiHow, http://fr.wikihow.com/créer-un-tableau-avec-Excel.

Paris-Dauphine University: Introduction to VBA for Excel, Philippe BERNARD, http://www.master272.com/finance/GP_L3/docs/VBA.pdf.

Towson University: Excel 2013: PivotTables and Macros, Pamela J.TAYLOR, https://www.towson.edu/adminfinance/ots/trainingdoc/customguide/Excel2013/ Excel%202013_PivotTables%20and%20Macros_studentFINAL.pdf.

College of Business Administration, Kansas State University: Conducting Data Analysis Using a Pivot Table, Brian KOVAR, http://info.cba.ksu.edu/bkovar/ PivotTableTutorial.pdf.

Tables in Excel 2007, http://silkyroad.developpez.com/excel/tableau/.

AgroParisTech: Manipulating lists of data with the Excel 2007 spreadsheet, Michel CARTEREAU, http://www.agroparistech.fr/mmip/mc/bazar/envoi.php? nom_fichier=cours-listes-excel2007.pdf.

Simulation

ExtendSim, simulation software, editor's website, http://www.extendsim.com.

Download demonstration version of ExtendSim, https://www.extendsim.com/prods_demo.html.

FlexSim, simulation software, editor's website, https://www.flexsim.com.

Arena, simulation software, editor's website, https://www.arenasimulation.com.

Witness, simulation software, editor's website, http://www.lanner.com.

SimWalk, pedestrian flow simulation software, editor's website, http://www.simwalk.com.

PathFinder, pedestrian movement and evacuation simulator, http://www.thunderheadeng.com/.

Official distributor of "ExtendSim" and "PathFinder" software, in France, http://www.1poin2.com.

Project Management

Sciforma, project management software, editor's website, http://www.sciforma.com.

Visual Planning, project management software, editor's website, http://www.visual-planning.com.

Microsoft Project, project management software, editor's website, https://products.office.com/fr-fr/project/project-and-portfolio-management-software.

GanttProject, project management software, editor's website, http://www.ganttproject.biz.

The wikipedia page for project management software, an enormous list of applications, http://fr.wikipedia.org/wiki/Logiciel_de_gestion_de_projets.

Index

Printed in the United States
By Bookmasters